# Lecture Notes in Mathematics

1611

Editors:
A. Dold, Heidelberg
F. Takens, Groningen

Springer
*Berlin*
*Heidelberg*
*New York*
*Barcelona*
*Budapest*
*Hong Kong*
*London*
*Milan*
*Paris*
*Santa Clara*
*Singapore*
*Tokyo*

Norbert Knarr

# Translation Planes

## Foundations and Construction Principles

Springer

Author

Norbert Knarr
Institut für Analysis
Technische Universität Braunschweig
Pockelsstr. 14
D-38106 Braunschweig, Germany

Cataloging-in-Publication Data applied for

Die Deutsche Bibliothek - CIP-Einheitsaufnahme

**Knarr, Norbert:**
Translation planes : foundations and construction principles /
Norbert Knarr. - Berlin ; Heidelberg ; New York : Springer,
1995
  (Lecture notes in mathematics ; 1611)
  ISBN 3-540-60208-9
NE: GT

Mathematics Subject Classification (1991): 51A40, 51H10, 51J15

ISBN 3-540-60208-9 Springer-Verlag Berlin Heidelberg New York

© Springer-Verlag Berlin Heidelberg 1995
Printed in Germany

Typesetting: Camera-ready TeX output by the author
SPIN: 10479552     46/3142-543210 – Printed on acid-free paper

# Table of Contents

# Introduction

An affine plane $\mathcal{A}$ is called a translation plane if the translation group of $\mathcal{A}$ operates transitively on the point set of $\mathcal{A}$. The fundamental results on translation planes were obtained by André in 1954. The translation group of $\mathcal{A}$ is isomorphic to the additive group of a vector space $V$ over a skewfield, and the points of $\mathcal{A}$ can be identified with the elements of $V$ in such a way that the lines of $\mathcal{A}$ are cosets of subspaces of $V$. The lines through the origin form a spread $\mathcal{B}$ of $V$, i.e. any two elements of $\mathcal{B}$ are complementary and the elements of $\mathcal{B}$ cover $V$. Hence, translation planes can be investigated using tools from linear algebra and projective geometry.

Primarily, we will be concerned with spreads of a 4-dimensional vector space. Equivalently, we can study systems of lines of a 3-dimensional projective space which are mutually disjoint and cover the space. These are also called spreads. First, we investigate spreads of 3-dimensional projective spaces over arbitrary skewfields. In later chapters we restrict our attention to topological spreads of real and complex projective spaces, for which our methods work especially well.

The first chapter contains an introduction to the theory of translation planes. We assume that the reader knows the basic facts about projective and affine planes. For arbitrary projective planes, the relevant definitions and theorems are given without proof. However, all results dealing directly with translation planes are proved explicitly.

Furthermore, we give a brief account of the theory of topological translation planes.

In the second chapter we discuss several possibilities for the description of spreads of 3-dimensional projective spaces.

Let $\mathcal{B}$ be a spread of a 3-dimensional projective space $\mathcal{P}$ and choose a line $S \in \mathcal{B}$. Let $E_1, E_2$ be distinct planes of $\mathcal{P}$ both of which contain $S$ and let $p$ be a point of $\mathcal{P}$ which is not contained in the union of $E_1$ and $E_2$. The affine plane obtained from $E_1$ by deleting the line $S$ is denoted by $E_1'$. With each line $G \in \mathcal{B} \setminus \{S\}$ we associate the point $G \cap E_1$ and the image of the point $G \cap E_2$ under the projection from $E_2$ to $E_1$ with center $p$. Since $\mathcal{B}$ is a spread, this defines a bijective mapping $f : E_1' \to E_1'$. We show that the mappings obtained in this way are generalizations of the transversal mappings invented by Ostrom for the description of spreads of finite projective spaces. If $\mathcal{B}$ is a dual spread instead of a spread, i.e. if every plane of $\mathcal{P}$ contains precisely one element of $\mathcal{B}$, then the same construction yields a mapping which is defined only on a subset of $E_1'$. It turns out that these mappings generalize the transversal mappings introduced by Betten for the study of topological spreads of the 3-dimensional real projective space. In order to distinguish them from the transversal mappings we call them *-transversal. We show that there exists a bijective correspondence between the set of all transversal or *-transversal mappings of an affine plane over a skewfield $F$ and the set of all spreads of a 3-dimensional projective space $\mathcal{P}$ over $F$ which contain a fixed line of $\mathcal{P}$.

If the skewfield $F$ admits an extension skewfield $L$ which has rank 2 as a right vector space over $F$, then the affine plane over $F$ can be identified with $L$. Moreover, the graphs of transversal mappings of the affine plane over $F$ can be viewed as subsets of the affine plane over $L$. According to Bruen, who studied this process for finite fields, we call the resulting sets indicator sets. We prove that a subset $\mathcal{J}$ of the affine plane over $L$ is an indicator set if and only if each line of the affine plane whose slope is contained in $F \cup \{\infty\}$ intersects $\mathcal{J}$ in precisely one point.

If $F$ is commutative and $L$ is a separable quadratic extension field of $F$, then we can associate an inversive plane $\Sigma(F, L)$ with the pair of fields $(F, L)$. The point set of $\Sigma(F, L)$ is the projective line $L \cup \{\infty\}$ and the circles of $\Sigma(F, L)$ are the images of $F \cup \{\infty\}$ under the group $\mathrm{PGL}_2(L)$. So the idea suggests itself to define also indicator sets with respect to other circles of $\Sigma(F, L)$. A natural candidate is the unit circle $L_1 = \{z \in L \mid z\bar{z} = 1\}$, where $^-$ denotes the involutorial $F$-automorphism of $L$. Indicator sets with respect to the unit circle are called $L_1$-indicator sets. Using a suitably defined Cayley transformation, we set up a bijective correspondence between the indicator sets and the $L_1$-indicator sets of the affine plane over $L$. If the 4-dimensional $F$-vector space underlying $\mathcal{P}$ is identified with $L \times L$, then almost all elements of $\mathcal{B}$ become graphs of linear mappings of the 2-dimensional $F$-vector space $L$. We show that algebraically $L_1$-indicator sets lead to the decomposition of these linear mappings into an $L$-linear and an $L$-antilinear part. This is similar to the Wirtinger calculus in complex analysis, where the real differential of a real differentiable mapping $f : \mathbb{C} \to \mathbb{C}$ is decomposed into its complex linear and its complex antilinear part.

In section 2.6 we examine the case $F = \mathbb{R}$ and $L = \mathbb{C}$ more closely. We show that $\mathbb{C}_1$-indicator sets in the complex affine plane can also be viewed as images of spreads of the 3-dimensional real projective space under the kinematic mapping of Blaschke and Grünwald. This connection is further investigated in chapter 3.

Since not every commutative field $F$ admits a separable quadratic extension field, in section 2.7 we replace the field $L$ by the ring $A = A(F)$ of double numbers over $F$. We introduce $A_1$-indicator sets and we show that every spread of the 4-dimensional $F$-vector space $A^2$ is associated with an $A_1$-indicator set.

In the third chapter we relate the theory developed in chapter 2 to the theory of kinematic spaces. Among other things we show that the fundamental properties of transversal mappings and of $L_1$- and $A_1$-indicator sets can also be obtained using this theory.

In chapter 4 we study the behaviour of $L_1$- and $A_1$-indicator sets under the application of linear mappings. Furthermore, we compute the quasifields associated with an $L_1$- or $A_1$-indicator set and we investigate the relation between algebraic properties of an $L_1$- or $A_1$-indicator set and geometric properties of the corresponding translation plane. In particular, we characterize the $L_1$- and $A_1$-indicator sets which lead to pappian planes and planes of Lenz type V.

In section 4.4 spreads covered by reguli are investigated. We introduce parabolic and hyperbolic flocks of reguli. Then we show that a spread $B$ contains a parabolic or hyperbolic flock of reguli if and only if the collineation group of the translation plane associated with $B$ contains a subgroup acting in a certain special way on $B$. This generalizes results which were obtained by Gevaert and Johnson for finite projective spaces.

Let $S$ be a line of a 3-dimensional projective space $P$ over a field $F$. It follows from the theory of $L_1$- and $A_1$-indicator sets that the reguli of $P$ which contain $S$ can be naturally identified with certain lines of a 4-dimensional affine space over $F$. Using this identification, we show that if a spread $B$ is covered by reguli all of which contain $S$, then $B$ contains either a hyperbolic or a parabolic flock of reguli. As a corollary, it follows that the translation plane associated with $B$ is pappian if and only if for each $Z, W \in B$ there exists a regulus $R$ of $P$ with $Z, W \in R$ and $R \subset B$.

In chapter 5 we study topological spreads of a 4-dimensional real vector space using $\mathbb{C}_1$-indicator sets. We show that every $\mathbb{C}_1$-indicator set is the graph of a contraction $\varphi : \mathbb{C} \to \mathbb{C}$ which in addition satisfies a condition at infinity. Furthermore, we derive necessary and sufficient conditions for the existence of parabolic or hyperbolic flocks of reguli in topological spreads. In particular, we prove that a topological spread $B$ of a 4-dimensional real vector space contains a parabolic flock of reguli if and only if the corresponding translation plane admits a 1-dimensional group of shears.

The locally compact 4-dimensional translation planes with an at least 7-dimensional collineation group were determined by Betten. Using the method of $\mathbb{C}_1$-indicator sets we derive simplified descriptions for some of these planes.

In chapter 6 we classify the locally compact planes of Lenz type V whose kernel is isomorphic to $\mathbb{C}$. Planes of this type are associated with topological spreads of a 4-dimensional complex vector space. We describe these planes using the double numbers over $\mathbb{C}$. This enables us to show that each of these planes can be obtained from two complex $2 \times 2$-matrices $B$ and $C$ which satisfy the condition $|m^t Bm| < |m^t Cm^*|$ for all $m \in \mathbb{C}^2 \setminus \{0\}$, where * denotes componentwise complex conjugation. It turns out that there exist two families of planes, which depend on either 1 or 3 real parameters. The 1-parameter family contains the planes over the division algebras of Rees; their collineation group is 17-dimensional. The 3-parameter family contains a 1-parameter subfamily of planes with an 18-dimensional collineation group and a 2-parameter subfamily with a 16-dimensional collineation group. The other planes of this family have a 15-dimensional collineation group. Up to now, only the locally compact 8-dimensional translation planes with an at least 17-dimensional collineation group had been classified by Hähl.

In chapter 7 we investigate topological spreads of 8- and 16-dimensional real vector spaces. Generalizing the results of chapter 5 we show that with every contraction of one of the normed real division algebras $\mathbb{H}$ or $\mathbb{O}$ which in addition satisfies a condition at infinity one can associate a topological spread of the real vector space $\mathbb{H}^2$ or $\mathbb{O}^2$, respectively. In contrast to the 4-dimensional case, not all topological spreads of these vector spaces can be obtained in this way. The locally compact 16-dimensional translation planes with an at least 38-dimensional collineation group were determined by Hähl. It turns out that these planes can be obtained from contractions of $\mathbb{O}$ which depend only on the real part or only on the absolute value of their argument.

# 1. Foundations

## 1.1 Translation Planes and Spreads

In this first section we give a comprehensive introduction to the theory of translation planes. For more detailed accounts the reader is referred to the original paper by André [1], or to the books by Lüneburg [74], Pickert [84] and Hughes-Piper [61].

We assume that the reader is familiar with the fundamentals of the theory of projective and affine planes. We shall use the following notation. $\mathcal{P} = (P, \mathcal{L})$ is a projective plane with point set $P$ and line set $\mathcal{L}$. The line joining two distinct points $p, q \in P$ is denoted by $p \vee q$. Dually, the intersection point of two distinct lines $L, M \in \mathcal{L}$ is denoted by $L \wedge M$. The collineation group of $\mathcal{P}$ is denoted by $\Sigma = \Sigma(\mathcal{P})$. For $p \in P$ the group of all collineations of $\mathcal{P}$ with center $p$ is defined by

$$\Sigma_{[p]} = \{\sigma \in \Sigma \mid \sigma(M) = M \text{ for all } M \in \mathcal{L} \text{ with } p \in M\}.$$

Dually, for $L \in \mathcal{L}$ the group of all collineations with axis $L$ is defined by

$$\Sigma_{[L]} = \{\sigma \in \Sigma \mid \sigma(q) = q \text{ for all } q \in P \text{ with } q \in L\}.$$

Furthermore, we put

$$\Sigma_{[p,L]} = \Sigma_{[p]} \cap \Sigma_{[L]}$$

for $p \in P$ and $L \in \mathcal{L}$. A collineation $\sigma$ has a center if and only if it has an axis. If the center is on the axis $\sigma$ is called an *elation*; otherwise $\sigma$ is a *homology*. The center and the axis of a non-identity central-axial collineation are unique. A central-axial collineation $\sigma \in \Sigma_{[p,L]}$ is determined by the image of one point of $\mathcal{P}$ which is not on $L$ and different from $p$.

If $\sigma \in \Sigma$ is a collineation of $\mathcal{P}$, then $\Sigma_{[p,L]}{}^{\sigma} = \sigma \Sigma_{[p,L]} \sigma^{-1} = \Sigma_{[\sigma(p),\sigma(L)]}$ for all $p \in P$ and $L \in \mathcal{L}$. For all $M, L \in \mathcal{L}$ the set $\Sigma_{[M,L]} = \bigcup_{p \in M} \Sigma_{[p,L]}$ is a group.

The group $\Sigma_{[p,L]}$ is called a *(linearly) transitive group of central-axial collineations* if $\Sigma_{[p,L]}$ acts transitively on the point sets $M \setminus \{p, L \wedge M\}$, where $M \in \mathcal{L}$ is a line incident with $p$. If $\Sigma_{[p,L]}$ is linearly transitive we also say that the plane $\mathcal{P}$ is *(p, L)-transitive*. The *Lenz-Barlotti figure* of $\mathcal{P}$ is defined by

$$\text{LBF}(\mathcal{P}) = \{(p, L) \in P \times \mathcal{L} \mid \mathcal{P} \text{ is } (p, L) - \text{transitive}\}.$$

Three points of $\mathcal{P}$ form a proper triangle if they are distinct and non-collinear. The point $p \in P$ is a center of the two proper triangles $p_1, p_2, p_3$ and $q_1, q_2, q_3$ if the lines $p_i \vee q_i$ contain the point $p$ for $i = 1, \ldots, 3$. Dually, $L \in \mathcal{L}$ is an axis of the proper triangles $p_1, p_2, p_3$ and $q_1, q_2, q_3$ if the points $r_i = (p_j \vee p_k) \wedge (q_j \vee q_k)$ are contained in $L$ for $\{i, j, k\} = \{1, 2, 3\}$. Let $p \in P, L \in \mathcal{L}$. We say that $\mathcal{P}$ is $(p, L)$-*desarguesian* if the following holds. Let $p_1, p_2, p_3$ and $q_1, q_2, q_3$ be proper triangles with center $p$ and assume that $r_1, r_2 \in L$. Then also $r_3 \in L$, i.e. $L$ is an axis of the triangles.

The plane $\mathcal{P}$ is $(p, L)$-transitive if and only if it is $(p, L)$-desarguesian.

If $\mathrm{LBF}(\mathcal{P}) = P \times \mathcal{L}$, i.e. $\mathcal{P}$ is $(p, L)$-desarguesian for all $(p, L) \in P \times \mathcal{L}$, then $\mathcal{P}$ is called desarguesian. The desarguesian planes are precisely the planes over skewfields.

If $\mathrm{LBF}(\mathcal{P}) = \{(p, L) \in P \times \mathcal{L} \mid p \in L\}$, then $\mathcal{P}$ is called a *Moufang plane*.

The plane $\mathcal{P}$ is called *pappian* if the following holds. Let $p_1, p_2, p_3$ and $q_1, q_2, q_3$ be triples of collinear points and put $r_i = (p_j \vee q_k) \wedge (q_j \vee p_k)$ for $\{i, j, k\} = \{1, 2, 3\}$. Then the points $r_1, r_2, r_3$ are collinear. The *Theorem of Hessenberg* says that every pappian plane is also desarguesian. Moreover, the pappian planes are precisely the planes over commutative fields.

Since we are usually dealing with affine planes, the following conventions turn out to be convenient. If $W$ is a line of an affine plane $\mathcal{A} = (P, \mathcal{L})$, then $w$ denotes the improper point of $W$, i.e. $W \cup \{w\}$ is a line of the projective extension of $\mathcal{A}$. The improper line of $\mathcal{A}$ is denoted $L_\infty$. The elements of $\Sigma_{[L_\infty]}$ are called *dilatations* of $\mathcal{A}$ and the dilatations whose center is on $L_\infty$ are called *translations*. The elations of $\mathcal{A}$ whose center is on $L_\infty$ but whose axis is different from $L_\infty$ are called *shears* of $\mathcal{A}$.

DEFINITION 1.1. Let $\mathcal{A} = (P, \mathcal{L})$ be an affine plane. $\mathcal{A}$ is called a *translation plane* if the translation group of $\mathcal{A}$ operates transitively on $P$.

Since a translation is determined by the image of one point, in fact the action is regular.

A projective plane $\mathcal{P} = (P, \mathcal{L})$ is called a translation plane if there exists a line $L \in \mathcal{L}$ such that the affine plane obtained from $\mathcal{P}$ by deleting $L$ is a translation plane. This is equivalent to $\{(p, L) \mid p \in L\} \subseteq \mathrm{LBF}(\mathcal{P})$.

LEMMA 1.2. *Let* $\mathcal{P} = (P, \mathcal{L})$ *be a projective plane, and let* $p, q \in P$ *and* $L, M \in \mathcal{L}$ *such that* $p \in M$ *and* $q \in L$. *Assume moreover that* $p \neq q$ *or* $L \neq M$. *Then the groups* $\Sigma_{[p, L]}$ *and* $\Sigma_{[q, M]}$ *centralize each other.*

*Proof.* Since $p \neq q$ or $L \neq M$ we have $\Sigma_{[p, L]} \cap \Sigma_{[q, M]} = \{1\}$. Let $\sigma \in \Sigma_{[p, L]}$ and $\tau \in \Sigma_{[q, M]}$. Then $\tau \sigma \tau^{-1} \in \Sigma_{[\tau(p), \tau(L)]} = \Sigma_{[p, L]}$ since $\tau(p) = p$ and $\tau(L) = L$. Hence $\tau \sigma \tau^{-1} \sigma^{-1} \in \Sigma_{[p, L]}$. Exchanging the roles of $\sigma$ and $\tau$ we get $\sigma \tau \sigma^{-1} \tau^{-1} \in \Sigma_{[q, M]}$ and hence $(\sigma \tau \sigma^{-1} \tau^{-1})^{-1} = \tau \sigma \tau^{-1} \sigma^{-1} \in \Sigma_{[p, L]} \cap \Sigma_{[q, M]} = \{1\}$. $\qquad\square$

COROLLARY 1.3. *The translation group of a translation plane is abelian.*

*Proof.* Let $\sigma, \tau \in \Sigma_{[L_\infty, L_\infty]} \setminus \{1\}$ be translations of $\mathcal{A}$. If $\sigma$ and $\tau$ have different centers they commute by Lemma 1.2. So we may assume that $\sigma$ and $\tau$ have the same center $p \in L_\infty$. Let $q \in L_\infty \setminus \{p\}$ and $\delta \in \Sigma_{[q, L_\infty]} \setminus \{1\}$, then the center of $\tau\delta$ is different from $p$ and $q$. Thus $\sigma(\tau\delta) = (\tau\delta)\sigma = \tau\sigma\delta$ by Lemma 1.2 and hence $\sigma\tau = \tau\sigma$. □

DEFINITION 1.4. Let $F$ be a skewfield and let $V$ be a vector space over $F$. A collection $\mathcal{B}$ of subspaces of $V$ with $|\mathcal{B}| \geq 3$ is called a *partial spread* of $V$ if the following condition is satisfied:

(P1) For any two different elements $U_1, U_2 \in \mathcal{B}$ we have $V = U_1 \oplus U_2$.

A partial spread is called a *spread* of $V$ if it also satisfies

(P2) Every vector $x \in V \setminus \{0\}$ is contained in an element of $\mathcal{B}$.

If $\mathcal{B}$ is a spread, the element of $\mathcal{B}$ whose existence is required by (P2) is uniquely determined by (P1).

The elements of a spread $\mathcal{B}$ are also called the components of $\mathcal{B}$.

THEOREM 1.5. *Let $\mathcal{B}$ be a spread of a vector space $V$ over a skewfield $F$. Put $P = V$ and $\mathcal{L} = \{U + x \,|\, U \in \mathcal{B}, x \in V\}$. Then $\mathcal{A} = \mathcal{A}(\mathcal{B}) = (P, \mathcal{L})$ is a translation plane. The translation group of $\mathcal{A}$ is isomorphic to $(V, +)$.*

*Proof.* We show first that $\mathcal{A}$ is an affine plane. Since $|F| \geq 2$ and $|\mathcal{B}| \geq 3$ every line of $\mathcal{A}$ contains at least 2 points and every point of $\mathcal{A}$ is on at least 3 lines. Let $x, y \in V$ be distinct points of $\mathcal{A}$. Then there exists precisely one element $U \in \mathcal{B}$ such that $x - y \in U$. Thus $U + x = U + y \in \mathcal{L}$ is the unique line of $\mathcal{A}$ connecting $x$ and $y$. Let $x \in V$ and $U + y \in \mathcal{L}$. The lines through $x$ are precisely the sets $W + x, W \in \mathcal{B}$. Such a line is parallel to $U + y$ if and only if $U = W$ since otherwise $U + W = V$. So there is a unique line through $x$ which is parallel to $U + y$. Hence $\mathcal{A}$ is an affine plane.

Obviously, the set $\{\tau_y : V \to V : x \mapsto x + y \,|\, y \in V\}$ is a transitive translation group of $\mathcal{A}$, and this group is isomorphic to $(V, +)$. □

It was proved by André [1] that the converse of Theorem 1.5 is also true, i.e. every translation plane can be obtained from a spread of a suitable vector space.

Let $\mathcal{A} = (P, \mathcal{L})$ be a translation plane. Since the translation group $\Sigma_{[L_\infty, L_\infty]}$ is abelian, it will be written additively.

LEMMA 1.6. *Let $p, q \in L_\infty$ be distinct points. Then $\Sigma_{[L_\infty, L_\infty]} = \Sigma_{[p, L_\infty]} + \Sigma_{[q, L_\infty]}$ and the sum is direct.*

*Proof.* As $\Sigma_{[p, L_\infty]} \cap \Sigma_{[q, L_\infty]} = \{0\}$ the sum is certainly direct. Since the action of $\Sigma_{[L_\infty, L_\infty]}$ on $P$ is regular, it is sufficient to show that $\Sigma_{[p, L_\infty]} + \Sigma_{[q, L_\infty]}$ acts transitively on $P$. Let $x, y \in P$. Put $z = (x \vee p) \wedge (y \vee q)$. Then there are

$\sigma \in \Sigma_{[p,L_\infty]}$ and $\tau \in \Sigma_{[q,L_\infty]}$ such that $\sigma(x) = z$ and $\tau(z) = y$. Consequently, $(\tau + \sigma)(x) = y$, and hence $\Sigma_{[p,L_\infty]} + \Sigma_{[q,L_\infty]}$ is transitive on $P$. □

We are now in position to prove André's representation theorem for translation planes.

**THEOREM 1.7.** *Let $\mathcal{A} = (P, \mathcal{L})$ be a translation plane. Put $V = \Sigma_{[L_\infty, L_\infty]}$ and $\mathcal{B} = \{\Sigma_{[p,L_\infty]} \mid p \in L_\infty\}$. Let the kernel of $\mathcal{A}$ be defined by $K(\mathcal{A}) = \{\delta \in \text{End}(V) \mid \delta(U) \subseteq U \text{ for all } U \in \mathcal{B}\}$, where $\text{End}(V)$ denotes the endomorphism ring of $V$. Then $K(\mathcal{A})$ is a skewfield, $V$ is a left vector space over $K(\mathcal{A})$ and $\mathcal{B}$ is a spread of $V$. Moreover, the translation planes $\mathcal{A}(\mathcal{B})$ and $\mathcal{A}$ are isomorphic.*

*Proof.* We show first that $K(\mathcal{A})$ is a ring. Let $\gamma, \delta \in K(\mathcal{A})$ and let $U \in \mathcal{B}$. Then $(\gamma - \delta)(U) \subseteq \gamma(U) + \delta(U) \subseteq U + U \subseteq U$ and $(\gamma \circ \delta)(U) = \gamma(\delta(U)) \subseteq \gamma(U) \subseteq U$. This shows that $K(\mathcal{A})$ is a subring of $\text{End}(V)$ and hence is a ring. Also, $V$ naturally is a left module over $K(\mathcal{A})$ and the elements of $\mathcal{B}$ are $K(\mathcal{A})$-submodules of $V$. Although we do not yet know that $K(\mathcal{A})$ is a skewfield, we can construct the incidence structure $\mathcal{A}(\mathcal{B})$ as in Theorem 1.5. From Lemma 1.6 we infer that $\mathcal{B}$ satisfies (P1) and since every translation has a center on $L_\infty$, condition (P2) is satisfied as well. Let $p \in P$ and define $\sigma : V \to P$ by $\sigma(\tau) = \tau(p)$. Then $\sigma$ is bijective since $V$ is sharply transitive on $P$. Moreover, $\sigma$ induces an isomorphism between $\mathcal{A}(\mathcal{B})$ and $\mathcal{A}$, as is easily seen. Hence $\mathcal{A}(\mathcal{B})$ is an affine plane.

So it remains to show that $K(\mathcal{A})$ is a skewfield.

Let $\delta \in K(\mathcal{A}) \setminus \{0\}$, where $0$ denotes the zero endomorphism. Assume that $\delta$ is not injective. Then there exists $x \in V \setminus \{0\}$ such that $\delta(x) = 0$. Let $U \in \mathcal{B}$ be the component with $x \in \mathcal{B}$. Let $y \in V \setminus U$ and let $W$ and $Z$ denote the unique elements of $\mathcal{B}$ for which $y \in W$ and $y + x \in Z$. Then $W$ and $Z$ are distinct and since $\delta(y) = \delta(y + x) \in \delta(W) \cap \delta(Z) \subseteq W \cap Z = \{0\}$, we get $\delta(y) = 0$. Applying this argument once more we conlude that $\delta$ is the zero endomorphism, contradicting our assumption. Hence $\delta$ is injective.

We show next that $\delta$ is also surjective. Let $x \in V \setminus \{0\}$ and let $U \in \mathcal{B}$ with $x \in U$. Choose $y \in V \setminus U$ and let $W \in \mathcal{B}$ with $y \in W$. Then $\delta(y) \neq x$ and hence there is a unique $Z \in \mathcal{B}$ with $\delta(y) - x \in Z$. Since $\mathcal{A}(\mathcal{B})$ is an affine plane and $U \neq Z$, there exists $z \in (Z + y) \cap U$. This implies $z - y \in Z$ and hence $\delta(z) - \delta(y) \in Z$. As $\delta(y) - x \in Z$ and $Z$ is a subgroup of $V$ we thus get $\delta(z) - x \in Z$. On the other hand, we have $z \in U$ and hence $\delta(z) \in U$. It follows that $\delta(z) - x \in U \cap Z = \{0\}$ and hence $\delta$ is surjective.

Let $U \in \mathcal{B}$, then $\delta^{-1}(U) = \delta^{-1}(\delta(U)) = U$. Thus $\delta^{-1} \in K(\mathcal{A})$ and $K(\mathcal{A})$ is a skewfield.

By definition of $K(\mathcal{A})$, the elements of $\mathcal{B}$ are $K(\mathcal{A})$-subspaces of $V$, hence $\mathcal{B}$ is a spread of $V$. □

Since we let our mappings operate from the left on their arguments, the translation group of a translation plane $\mathcal{A}$ naturally becomes a left vector space over the kernel of $\mathcal{A}$. It is also possible to use a right vector space for the representation of $\mathcal{A}$. To this end we replace the skewfield $(K(\mathcal{A}), +, \cdot)$ by its *opposite skewfield* $(\widetilde{K(\mathcal{A})}, \tilde{+}, \tilde{\cdot})$, where $a\tilde{+}b = a+b$ and $a\tilde{\cdot}b = b \cdot a$ for $a, b \in K(\mathcal{A})$. Every left vector space $V$ over $K(\mathcal{A})$ becomes a right vector space over $\widetilde{K(\mathcal{A})}$ if we define $x\tilde{\cdot}c = c \cdot x$ for $x \in V, c \in \widetilde{K(\mathcal{A})} = K(\mathcal{A})$. The subspaces of the left $K(\mathcal{A})$-vector space $V$ coincide with the subspaces of the right $\widetilde{K(\mathcal{A})}$-vector space $V$. Hence, $\mathcal{B}$ is also a spread of the right $\widetilde{K(\mathcal{A})}$-vector space $V$.

DEFINITION 1.8. Let $\mathcal{A}$ be a translation plane and let $F$ be a skewfield. We say that $\mathcal{A}$ admits a *representation* over $F$ if there exists a vector space $V$ over $F$ and a spread $\mathcal{B}$ of $V$ such that $\mathcal{A}$ is isomorphic to $\mathcal{A}(\mathcal{B})$.

PROPOSITION 1.9. *Let $\mathcal{A}$ be a translation plane and let $F$ be a skewfield. Then $\mathcal{A}$ admits a representation over $F$ if and only if $F$ is isomorphic or antiisomorphic to a subskewfield of $K(\mathcal{A})$. More precisely: $\mathcal{A}$ admits a representation in a left (right) vector space over $F$ if and only if $F$ is isomorphic (antiisomorphic) to a subskewfield of $K(\mathcal{A})$. In particular, $K(\mathcal{A})$ and $\widetilde{K(\mathcal{A})}$ are the largest skewfields over which $\mathcal{A}$ admits a representation.*

*Proof.* Let $F$ be a subskewfield of $K(\mathcal{A})$. Since every left vector space over $K(\mathcal{A})$ also is a left vector space over $F$, the translation plane $\mathcal{A}$ admits a representation over $F$ and hence also over any skewfield isomorphic to $F$. A similar argument applies to subskewfields of $\widetilde{K(\mathcal{A})}$.

Assume now that $\mathcal{A}$ admits a representation over a skewfield $F$. Let $V$ be a vector space over $F$ and let $\mathcal{B}$ be a spread of $V$ such that $\mathcal{A}$ is isomorphic to $\mathcal{A}(\mathcal{B})$.

If $V$ is a left vector space we define $K' = \{\delta_c : V \to V : x \mapsto cx \mid c \in F\}$. Then $F'$ is a subskewfield of $K(\mathcal{A}(\mathcal{B}))$ which is isomorphic to $F$. Hence $K(\mathcal{A})$ contains a subskewfield which is isomorphic to $F$.

If $V$ is a right vector space we define instead $K' = \{\delta_c : V \to V : x \mapsto xc \mid c \in F\}$. Then $F'$ is a subskewfield of $K(\mathcal{A}(\mathcal{B}))$ which is antiisomorphic to $F$. Hence $K(\mathcal{A})$ contains a subskewfield which is antiisomorphic to $F$. $\square$

If $K(\mathcal{A})$ is a field, the distinction between left and right vector spaces vanishes. By Wedderburn's theorem, cf. e.g. [62: p.453], every finite skewfield is commutative, hence the kernel of a finite translation plane is a field. The same holds for locally compact connected translation planes with the exception of the quaternion plane, cf. Proposition 1.29.

Let $\mathcal{P}$ be the projective space associated with the vector space $V$ and let $\mathcal{B}$ be a spread of $V$. Viewed projectively, $\mathcal{B}$ has the following properties:

(S1) Any two distinct subspaces $U_1, U_2 \in \mathcal{B}$ intersect trivially and span $\mathcal{P}$.

(S2) Every point of $\mathcal{P}$ is contained in an element of $\mathcal{B}$.

A system of subspaces of $\mathcal{P}$ satisfying (S1) and (S2) will be called a *spread* of the projective space $\mathcal{P}$.

## 1.2 Quasifields and Spread Sets

Let $\mathcal{B}$ be a spread of the $F$-vector space $V$ and let $W, S \in \mathcal{B}$ be distinct. Then $V$ is the direct sum of $W$ and $S$ and for every $U \in \mathcal{B} \setminus \{S\}$ we have $U \cap S = \{0\}$. Hence, every component $U \in \mathcal{B} \setminus \{S\}$ is the graph of a linear mapping $\lambda_U : W \to S$. In particular, $\lambda_W$ is the zero mapping. It follows easily from (P1) and (P2) that the set $\mathcal{M} = \{\lambda_U : W \to S \mid U \in \mathcal{B} \setminus \{S\}\}$ has the following characteristic properties:

(L1) For all $U, Z \in \mathcal{B} \setminus \{S\}$ with $U \neq Z$ the mapping $\lambda_U - \lambda_Z$ is bijective.

(L2) For all $x \in W \setminus \{0\}$ the mapping $\varrho_x : \mathcal{B} \setminus \{S\} \to S : U \mapsto \lambda_U(x)$ is surjective.

Since the vector spaces $W$ and $S$ are isomorphic, they usually are identified. This motivates the following

DEFINITION 1.10. Let $X$ be a vector space over a skewfield $F$. A collection of linear mappings $\mathcal{M} \subseteq \operatorname{End}_F(X)$ is called a *spread set* of $X$ if the following conditions are satisfied:

(M1) For any two distinct elements $\lambda_1, \lambda_2 \in \mathcal{M}$ the mapping $\lambda_1 - \lambda_2$ is bijective.

(M2) For all $x \in X \setminus \{0\}$ the mapping $\varrho_x : \mathcal{M} \to X : \lambda \mapsto \lambda(x)$ is surjective.

It follows from (M1) that the mappings $\varrho_x$ considered in (M2) are injective for $x \in X \setminus \{0\}$. Hence, if (M2) is satisfied, they are even bijective.

Actually, it is sufficient to require that $X$ is an abelian group instead of a vector space and that $\mathcal{M} \subseteq \operatorname{End}(X)$ satisfies (M1) and (M2). It then follows from Theorem 1.7 that $X$ is a vector space over a suitable skewfield and that the elements of $\mathcal{M}$ are linear mappings.

PROPOSITION 1.11. *Let $X \neq \{0\}$ be a vector space over a skewfield $F$ and let $\mathcal{M} \subset \operatorname{End}_F(X)$ be a spread set of $X$. Put $V = X \times X$ and $S = \{0\} \times X$. For $\lambda \in \mathcal{M}$ let $U_\lambda = \{(x, \lambda(x)) \mid x \in X\}$ denote the graph of $\lambda$. Then $\mathcal{B}(\mathcal{M}) = \{S\} \cup \{U_\lambda \mid \lambda \in \mathcal{M}\}$ is a spread of $V$. Conversely, every spread $\mathcal{B}$ of $V$ with $S \in \mathcal{B}$ is obtained from a spread set of $X$ in the way just described.*

*Proof.* Let $\mathcal{M}$ be a spread set of $X$ and define $\mathcal{B}$ as in the proposition.

Let $\lambda, \mu \in \mathcal{M}$ be distinct. We have to show that $V = U_\lambda \oplus U_\mu$. Let $(w, z) \in V$. Then we have

$$
\begin{aligned}
(w, z) &= (x, \lambda(x)) + (y, \mu(y)) \\
&= (x + y, \lambda(x) + \mu(y)) \\
&= (w, \lambda(x) + \mu(w - x)) \\
&= (w, \lambda(x) - \mu(x) + \mu(w)).
\end{aligned}
$$

Since $\lambda - \mu$ is bijective, this equation has a unique solution $x \in X$. So $y \in X$ is determined uniquely as well and hence $V = U_\lambda \oplus U_\mu$.

Let $(w, z) \in V \setminus \{(0,0)\}$. If $w = 0$ then $(w, z) \in S$. So assume $w \neq 0$. We need to find $\lambda \in \mathcal{M}$ such that $(w, z) \in U_\lambda$. The equation

$$(w, z) = (x, \lambda(x)) = (w, \lambda(w)) = (w, \varrho_w(\lambda))$$

has a solution $\lambda \in \mathcal{M}$ since $\varrho_w$ is surjective. So $(w, z)$ is contained in an element of $\mathcal{B}$ and hence $\mathcal{B}$ is a spread of $V$.

Assume now that $\mathcal{B}$ is a spread of $V$ with $S \in \mathcal{B}$. Let $U \in \mathcal{B} \setminus \{S\}$. Since $V = X \times X$ and $U \cap S = \{(0,0)\}$, there exists a linear mapping $\lambda : X \to X$ such that $U = U_\lambda$. By reversing the arguments given above it is easily seen that $\mathcal{M} = \{\lambda \in \mathrm{End}_F(X) \,|\, U_\lambda \in \mathcal{B} \setminus \{S\}\}$ is a spread set of $X$, and obviously we have $\mathcal{B}(\mathcal{M}) = \mathcal{B}$. $\qquad\square$

It follows from elementary linear algebra that if a vector space $V$ contains a spread $\mathcal{B}$ then there exists a vector space $X$ such that $V$ can be identified with $X \times X$ and $S = \{0\} \times X \in \mathcal{B}$. Hence, every spread can be obtained from a suitable spread set.

DEFINITION 1.12. Let $Q$ be a set equipped with two binary operations $+, \circ : Q \times Q \to Q$. For $a \in Q$ we define the mappings $\lambda_a, \varrho_a : Q \to Q$ by $\lambda_a(x) = a \circ x$ and $\varrho_a(x) = x \circ a$, respectively. Then $(Q, +, \circ)$ is called a *right quasifield* if the following axioms are satisfied:

(Q1) $(Q, +)$ is an abelian group.

(Q2) $x \circ 0 = 0 \circ x = 0$ for all $x \in Q$.

(Q3) There exists an element $1 \in Q \setminus \{0\}$ such that $1 \circ x = x \circ 1 = x$ for all $x \in Q$.

(Q4) For all $m, x, y \in Q$ we have $(x + y) \circ m = x \circ m + y \circ m$.

(Q5) For all $m, n \in Q$ with $m \neq n$ the mapping $\varrho_m - \varrho_n : Q \to Q : x \mapsto x \circ m - x \circ n$ is bijective.

(Q6) For all $x \in Q \setminus \{0\}$ the mapping $\lambda_x : Q \to Q : m \mapsto x \circ m$ is surjective.

The *kernel* of $Q$ is defined by $K(Q) = \{k \in Q \,|\, k \circ (x + y) = k \circ x + k \circ y$ and $k \circ (x \circ y) = (k \circ x) \circ y$ for all $x, y \in Q\}$.

The axioms for a left quasifield are obtained from (Q1) - (Q6) by exchanging the factors in all products that appear.

It follows from (Q5) that the mappings $\lambda_x$ are injective for $x \in Q \setminus \{0\}$. Hence, they are even bijective if (Q6) is satisfied.

LEMMA 1.13. *Let $Q$ be a right quasifield with kernel $K(Q)$. Then $K(Q)$ is a skewfield and $Q$ is a left vector space over $K(Q)$. Moreover, the mappings $\varrho_m : Q \to Q : x \mapsto x \circ m$ are linear over $K(Q)$ for all $m \in Q$.*

*Proof.* Let $k, l, x \in Q$ with $k + l = 0$. Then we have $k \circ x + l \circ x = (k+l) \circ x = 0 \circ x = 0$. It follows that $-(k \circ x) = (-k) \circ x$.

Let $k, l \in K(Q)$ and $x, y \in Q$. Then we have

$$(k - l) \circ (x + y) = k \circ (x + y) - l \circ (x + y)$$
$$= k \circ x + k \circ y - l \circ x - l \circ y$$
$$= (k - l) \circ x + (k - l) \circ y.$$

Also we have

$$(k - l) \circ (x \circ y) = k \circ (x \circ y) - l \circ (x \circ y)$$
$$= (k \circ x) \circ y - (l \circ x) \circ y$$
$$= (k \circ x - l \circ x) \circ y$$
$$= ((k - l) \circ x) \circ y.$$

This shows $(k - l) \in K(Q)$. By a similar argument, also $k \circ l \in K(Q)$.

Let $x, y \in Q$. From (Q3) we infer that $1 \circ (x + y) = x + y = (1 \circ x) + (1 \circ y)$ and $1 \circ (x \circ y) = x \circ y = (1 \circ x) \circ y$. It follows that $1 \in K(Q)$.

By definition of $K(Q)$, the multiplication of $Q$ restricted to $K(Q)$ is associative, Moreover, $K(Q)$ satisfies both distributive laws. Hence, $K(Q)$ is a ring.

Let $k \in K(Q) \setminus \{0\}$. By (Q6) there exists $l \in Q$ such that $k \circ l = 1$. Let $x, y \in Q$. Then we get

$$k \circ (l \circ (x \circ y)) = (k \circ l) \circ (x \circ y) = (x \circ y)$$
$$= ((k \circ l) \circ x) \circ y = (k \circ (l \circ x)) \circ y = k \circ ((l \circ x) \circ y).$$

It follows that $lo(xoy) = (lox)oy$. By a similar argument also $lo(x+y) = lox+loy$ and hence $l \in K(Q)$. Thus $K(Q)$ is a skewfield.

It follows from the definition of $K(Q)$ that $Q$ is a left vector space over $K(Q)$.

From (Q4) we infer that the mappings $\varrho_m : Q \to Q : x \mapsto x \circ m$ are additive for all $m \in Q$. Let $x, m \in Q$ and $k \in K(Q)$. Then we have

$$\varrho_m(k \circ x) = (k \circ x) \circ m = k \circ (x \circ m) = k \circ \varrho_m(x).$$

Hence $\varrho_m$ is linear for all $m \in Q$. $\square$

**PROPOSITION 1.14.** *Let $Q$ be a right quasifield and define $\mathcal{M} = \mathcal{M}(Q) = \{\varrho_m : Q \to Q \mid m \in Q\}$. Then $\mathcal{M}$ is a spread set of the $K(Q)$-vector space $Q$. Let $\mathcal{A} = \mathcal{A}(Q) = \mathcal{A}(B(\mathcal{M}(Q)))$ denote the translation plane associated with the spread $B(\mathcal{M}(Q))$. Then we have $K(\mathcal{A}) = \{\delta_k : Q^2 \to Q^2 : (x, y) \mapsto (k \circ x, k \circ y) \mid k \in K(Q)\}$. In particular, the skewfields $K(Q)$ and $K(\mathcal{A})$ are isomorphic.*

*Proof.* By Lemma 1.13, the elements of $\mathcal{M}$ are linear mappings of the left $K(Q)$-vector space $Q$.

Obviously, (Q5) is equivalent to (M1) and (Q6) is equivalent to (M2). Hence $\mathcal{M}$ is a spread set.

Let $\mathcal{A} = \mathcal{A}(Q)$ be the translation plane associated with $Q$.

Let $k \in K(Q)$ and consider the mapping $\delta_k : Q^2 \to Q^2 : (x, y) \mapsto (k \circ x, k \circ y)$. Then $\delta_k(S) = S$, where $S = \{0\} \times Q$. Let $m \in Q$ and let $U_m \in \mathcal{B}(\mathcal{M}(Q))$ be the component associated with $m$, i.e. $U_m = \{(x, x \circ m) \mid x \in Q\}$. Then we have

$$\delta_k(U_m) = \{(k \circ x, k \circ (x \circ m)) \mid x \in Q\} = \{(k \circ x, (k \circ x) \circ m) \mid x \in Q\} \subseteq U_m$$

since $k \in K(Q)$. It follows that $\delta_k \in K(\mathcal{A})$.

Let now $\delta : Q^2 \to Q^2$ be an element of $K(\mathcal{A})$. Then $\delta$ is an endomorphism of the additive group $(Q^2, +)$. Since $\delta(S) \subseteq S$ and $\delta(U_0) \subseteq U_0$, there exist additive endomorphisms $\kappa, \kappa' : Q \to Q$ such that $\delta(x, y) = (\kappa(x), \kappa'(y))$ for all $x, y \in Q$. Now we have

$$\delta(U_1) = \{(\kappa(x), \kappa'(x)) \mid x \in Q\} \subseteq U_1 = \{(x, x) \mid x \in Q\}$$

and hence $\kappa = \kappa'$.

Let $m \in Q$. Then we have

$$\delta(U_m) = \{(\kappa(x), \kappa(x \circ m)) \mid x \in Q\} \subseteq U_m = \{(x, x \circ m) \mid x \in Q\}.$$

It follows that $\kappa(x \circ m) = \kappa(x) \circ m$ for all $x, m \in Q$. We now define $k = \kappa(1)$, then we have $\kappa(x) = k \circ x$. Let $x, m \in Q$. Then we compute

$$k \circ (x \circ m) = \kappa(x \circ m) = \kappa(x) \circ m = (k \circ x) \circ m,$$

which implies $k \in K(Q)$. $\qquad\qquad\square$

The lines of $\mathcal{A}$ which are not parallel to $S = \{0\} \times Q$ are given by equations of the form $y = x \circ m + b$. We call $m$ the *slope* and $b$ the *y-intercept* of the respective line.

Every left quasifield $Q$ is a right vector space over its kernel and the kernel of the translation plane $\mathcal{A}(Q)$ is antiisomorphic to the kernel of $Q$. The lines are given by equations of the form $y = m \circ x + b$.

Our next aim is to show that in fact every translation plane can be coordinatized by a suitable quasifield. Since every translation plane is obtained from a spread set, it is sufficient to show that every spread set is associated with a quasifield. Since a quasifield contains 0 and 1, the associated spread set contains the zero endomorphism and the identity mapping. This is not true for arbitrary spread sets and so we are essentially dealing with a normalization problem. A spread set $\mathcal{M}$ of a vector space $X$ over a skewfield $F$ is called *normalized* if $\mathcal{M}$ contains the zero and the identity endomorphism of $X$.

LEMMA 1.15.   *Let $X$ be a left vector space over a skewfield $F$ and let $\mathcal{M} \subseteq \mathrm{End}_F(X)$ be a spread set of $X$. Let $\lambda_0, \lambda_1 \in \mathcal{M}$ be distinct. Put $\mathcal{M}' = \{(\lambda - \lambda_0)(\lambda_1 - \lambda_0)^{-1} \mid \lambda \in \mathcal{M}\}$. Then $\mathcal{M}'$ is a normalized spread set of $X$ and the translation planes $\mathcal{A}(\mathcal{B}(\mathcal{M}))$ and $\mathcal{A}(\mathcal{B}(\mathcal{M}'))$ are isomorphic.*

*Proof.* Let the linear mapping $\varphi : X \times X \to X \times X$ be defined by

$$\varphi(x, y) = ((\lambda_1 - \lambda_0)(x), y - \lambda_0(x)).$$

For $\lambda \in \mathcal{M}$ let $U_\lambda = \{(x, \lambda(x)) \mid x \in X\}$ denote the graph of $\lambda$. Then we have

$$\varphi(U_\lambda) = \{((\lambda_1 - \lambda_0)(x), \lambda(x) - \lambda_0(x)) \mid x \in X\}$$
$$= \{(x, (\lambda - \lambda_0)(\lambda_1 - \lambda_0)^{-1}(x) \mid x \in X\} = U_{(\lambda - \lambda_0)(\lambda_1 - \lambda_0)^{-1}}.$$

It follows that $\mathcal{M}'$ is a spread set and that $\varphi$ induces an isomorphism from $\mathcal{A}(\mathcal{B}(\mathcal{M}))$ to $\mathcal{A}(\mathcal{B}(\mathcal{M}'))$, cf. also Theorem 1.18.

By setting $\lambda = \lambda_0$ and $\lambda = \lambda_1$ we see that $\mathcal{M}'$ contains the zero and the identity endomorphism and hence is normalized. $\quad\square$

PROPOSITION 1.16. *Let $\mathcal{A}$ be a translation plane. Then there exists a quasifield $Q$ such that $\mathcal{A}$ and $\mathcal{A}(Q)$ are isomorphic.*

*Proof.* By Theorem 1.7 and Proposition 1.11 we may assume that $\mathcal{A}$ is constructed from a spread set $\mathcal{M}$ of a left vector space $X$ over a skewfield $F$. Because of Lemma 1.15 we may further assume that $\mathcal{M}$ contains the zero and the identity endomorphism.

Let $e \in X \setminus \{0\}$ and let $\varrho_e : \mathcal{M} \rightarrow X$ be defined by $\varrho_e(\lambda) = \lambda(e)$ for $\lambda \in \mathcal{M}$. By (M2) $\varrho_e$ is bijective. We define a multiplication $\circ : X \times X \rightarrow X$ by $x \circ m = (\varrho_e^{-1}(m))(x)$ for $x, m \in X$.

We claim that $(X, +, \circ)$ is a right quasifield.

As $(X, +)$ is the additive group of a vector space, (Q1) is certainly satisfied.

Let $x \in X$, then we have $0 \circ x = (\varrho_e^{-1}(x))(0) = 0$ since $(\varrho_e^{-1})(x)$ is a linear mapping. Let $\lambda_0 : X \rightarrow X$ be the zero endomorphism. Then $\varrho_e(\lambda_0) = \lambda_0(e) = 0$ and hence $\varrho_e^{-1}(0) = \lambda_0$. Consequently, $x \circ 0 = (\varrho_e^{-1}(0))(x) = \lambda_0(x) = 0$. Thus (Q2) is satisfied.

Let $\lambda_1 : X \rightarrow X$ be the identity, then we have $\varrho_e(\lambda_1) = \lambda_1(e) = e$.

Let $x \in X$, then $x \circ e = (\varrho_e^{-1}(e))(x) = \lambda_1(x) = x$. Put $\lambda = \varrho_e^{-1}(x)$, then $x = \varrho_e(\lambda) = \lambda(e)$. This yields

$$e \circ x = (\varrho_e^{-1}(x))(e) = (\varrho_e^{-1}(\varrho_e(\lambda)))(e) = \lambda(e) = x.$$

Hence $e$ is a multplicative identity and (Q3) is satisfied.

Since the elements of $\mathcal{M}$ are linear mappings of $X$, axiom (Q4) is satisfied.

The properties (Q5) and (Q6) are equivalent to (M1) and (M2), respectively.

It follows that $(X, +, \circ)$ is a right quasifield and by construction $\mathcal{M}(X) = \mathcal{M}$. $\quad\square$

If $\mathcal{M}$ is a spread set of a right vector space $X$ we define accordingly $m \circ x = (\varrho_e^{-1}(m))(x)$ for $m, x \in X$, where $e \in X \setminus \{0\}$. Then $(X, +, \circ)$ is a left quasifield which coordinatizes the translation plane associated with $\mathcal{M}$.

In general, left vector spaces correspond to right quasifields and right vector spaces correspond to left quasifields.

Although we have developed the theory for left vector spaces and right quasifields, in later chapters we will normally use right vector spaces and left quasifields.

## 1.3 Isomorphisms and Collineations of Translation Planes

In this section we study isomorphisms and collineations of translation planes.

LEMMA 1.17. *Let* $\mathcal{A} = (P, \mathcal{L})$ *be a translation plane with kernel* $K(\mathcal{A})$. *Put* $V = \Sigma_{[L_\infty, L_\infty]}$ *and* $\mathcal{B} = \{\Sigma_{[p, L_\infty]} \mid p \in L_\infty\}$. *For* $p \in P$ *let* $\sigma : V \to P$ *be defined by* $\sigma(\tau) = \tau(p)$. *Then* $\Sigma_{[p, L_\infty]} = \{\sigma\delta\sigma^{-1} \mid \delta \in K(\mathcal{A})\backslash\{0\}\}$. *In particular,* $\Sigma_{[p, L_\infty]}$ *is isomorphic to the multiplicative group of the skewfield* $K(\mathcal{A})$.

*Proof.* By Theorem 1.7, $\mathcal{B}$ is a spread of the left $K(\mathcal{A})$-vector space $V$ and $\sigma$ induces an isomorphism between the translation planes $\mathcal{A}(\mathcal{B})$ and $\mathcal{A}$. Hence we may assume that $P = V$, $p = 0$ and $\sigma = \mathrm{id}$.

Let $\delta \in \Sigma_{[p, L_\infty]}$. Since $V = \Sigma_{[L_\infty, L_\infty]}$ is a normal subgroup of the collineation group of $\mathcal{A}$, we have $\delta \in \mathrm{Aut}(V) \subseteq \mathrm{End}(V)$, where $\mathrm{Aut}(V)$ denotes the automorphism group of $V$. The lines through $p = 0$ are precisely the elements of $\mathcal{B}$ and since $\delta$ fixes all these lines, we get $\delta \in K(\mathcal{A}) \backslash \{0\}$.

Let now $\delta \in K(\mathcal{A}) \backslash \{0\}$, then $\delta$ is bijective since $K(\mathcal{A})$ is a skewfield. By definition, $\delta$ fixes the elements of $\mathcal{B}$. Hence, $\delta$ permutes the cosets of these elements and so is a collineation of $\mathcal{A}$. Obviously, $\delta \in \Sigma_{[p, L_\infty]}$. $\square$

Because of Lemma 1.17, the elements of $\Sigma_{[p, L_\infty]}$ are also called *kernel homologies* of $\mathcal{A}$.

A simple consequence of Lemma 1.17 is the following fundamental theorem for translation planes, which is also due to André [1].

THEOREM 1.18. *For* $i = 1, 2$ *let* $V_i$ *be a left vector space over a skewfield* $F_i$ *and let* $\mathcal{B}_i$ *be a spread of* $V_i$. *Assume moreover that* $F_i$ *is the kernel of the translation plane* $\mathcal{A}_i = \mathcal{A}(\mathcal{B}_i)$. *Let* $\sigma : V_1 \to V_2$ *be bijective. Then* $\sigma$ *induces an isomorphism between* $\mathcal{A}_1$ *and* $\mathcal{A}_2$ *if and only if there exists a semilinear mapping* $\lambda : V_1 \to V_2$ *and a vector* $t \in V_2$ *such that* $\lambda(\mathcal{B}_1) = \mathcal{B}_2$ *and* $\sigma(x) = \lambda(x) + t$ *for all* $x \in V_1$. *In particular, the collineation group of an affine translation plane is isomorphic to the semidirect product of the translation group with the group of all mappings which are semilinear over the kernel and fix the spread.*

*Proof.* Assume first that there exists a semilinear mapping $\lambda : V_1 \to V_2$ and a vector $t \in V_2$ such that $\lambda(\mathcal{B}_1) = \mathcal{B}_2$ and $\sigma(x) = \lambda(x) + t$ for all $x \in V_1$.

Let $U \in \mathcal{B}_1$ and $x \in V_1$. Then $\sigma(U + x) = \lambda(U + x) + t = \lambda(U) + \lambda(x) + t$. Since $\lambda(U) \in \mathcal{B}_2$, the mapping $\sigma$ induces an isomorphism between $\mathcal{A}_1$ and $\mathcal{A}_2$.

Assume now that $\sigma$ induces an isomorphism between $\mathcal{A}_1$ and $\mathcal{A}_2$. Put $t = \sigma(0)$ and define $\lambda : V_1 \to V_2$ by $\lambda(x) = \sigma(x) - t$. Then $\lambda(0) = 0$ and $\lambda$ also induces an isomorphism between $\mathcal{A}_1$ and $\mathcal{A}_2$. Since $\lambda$ maps the lines through $0 \in V_1$ to the lines through $0 \in V_2$, we have $\lambda(\mathcal{B}_1) = \mathcal{B}_2$. So it remains to show that $\lambda$ is semilinear.

Let $x \in V_1$ and let $\tau_x : V_1 \to V_1 : y \mapsto x + y$ denote the translation associated with $x$. Then $\lambda\tau_x\lambda^{-1} : V_2 \to V_2$ is a translation of $\mathcal{A}_2$. Since $\lambda\tau_x\lambda^{-1}(0) =$

$\lambda \tau_x(0) = \lambda(x)$, we get $\lambda \tau_x \lambda^{-1} = \tau_{\lambda(x)}$ and hence $\lambda \tau_x = \tau_{\lambda(x)} \lambda$. Let $y \in V_1$. Then we have

$$\lambda(x + y) = \lambda(\tau_x(y)) = \tau_{\lambda(x)}(\lambda(y)) = \lambda(x) + \lambda(y)$$

and hence $\lambda$ is additive.

Let $a \in F_1$ and let $\delta_a : V_1 \to V_1$ be defined by $\delta_a(x) = ax$. Since $F_1$ is the kernel of $\mathcal{A}_1$, the mappings $\delta_a, a \in F_1^\times$, are precisely the homologies of $\mathcal{A}_1$ with center $0$ and axis $L_\infty$. Since $\lambda(0) = 0$, the mapping $\lambda \delta_a \lambda^{-1} : V_2 \to V_2$ is a homology of $\mathcal{A}_2$ with center $0$ and axis $L_\infty$ for all $a \in F_1^\times$. Thus there exists an injective mapping $\alpha : F_1 \to F_2$ such that $\lambda \delta_a \lambda^{-1} = \delta_{\alpha(a)}$ for all $a \in F_1$. Since $F_2$ is the kernel of $\mathcal{A}_2$, the mapping $\alpha$ is also surjective and hence bijective. We show that $\alpha$ is an isomorphism of skewfields.

Let $a, b \in F_1$. Then we have

$$\delta_{\alpha(a+b)} = \lambda \delta_{(a+b)} \lambda^{-1} = \lambda(\delta_a + \delta_b)\lambda^{-1}$$
$$= \lambda \delta_a \lambda^{-1} + \lambda \delta_b \lambda^{-1} = \delta_{\alpha(a)} + \delta_{\alpha(b)} = \delta_{\alpha(a)+\alpha(b)}$$

and hence $\alpha(a + b) = \alpha(a) + \alpha(b)$. By a similar argument also $\alpha(ab) = \alpha(a)\alpha(b)$. It follows that $\alpha$ is an isomorphism of skewfields.

Let $x \in V_1$ and $c \in F_1$. Then we have

$$\lambda(cx) = \lambda(\delta_c(x)) = \delta_{\alpha(c)}(\lambda(x)) = \alpha(c)\lambda(x).$$

Hence, $\lambda$ is semilinear with companion isomorphism $\alpha$. $\square$

If we drop the assumption that $F_i$ is the kernel of $\mathcal{A}_i$ the theorem does not hold any longer. So, for example, the complex affine plane admits collineations which are not real linear [67].

Let $\mathcal{B}_1$ and $\mathcal{B}_2$ be spreads of vector spaces $V_1$ and $V_2$, respectively. We say that $\mathcal{B}_1$ and $\mathcal{B}_2$ are *equivalent* if the translation planes $\mathcal{A}(\mathcal{B}_1)$ and $\mathcal{A}(\mathcal{B}_2)$ are isomorphic. Accordingly, two spread sets $\mathcal{M}_1$ and $\mathcal{M}_2$ are called *equivalent* if the spreads $\mathcal{B}(\mathcal{M}_1)$ and $\mathcal{B}(\mathcal{M}_2)$ are equivalent. For finite dimensional vector spaces over finite prime fields the problem to determine whether two spread sets are equivalent or not was treated by Maduram [78].

Up to now we were only considering affine planes, but Theorem 1.18 also allows us to determine the collineations of a projective translation plane. If the projective extension $\mathcal{P}$ of an affine translation plane $\mathcal{A}$ admits a collineation which moves the improper line, then $\mathcal{P}$ is a Moufang plane by the theorem of Skornjakov-San Soucie [61: Theorem 6.16]. The collineation group of a Moufang plane acts transitively on the set of lines.

Let $Q$ be a right quasifield and let $\mathcal{A}$ denote the translation plane associated with $Q$. From Proposition 1.14 and Lemma 1.17 we infer that the group of homologies of $\mathcal{A}$ with center $(0,0)$ and axis $L_\infty$ consists of all mappings $Q^2 \to Q^2 : (x,y) \mapsto (k \circ x, k \circ y)$, where $k \in K(Q) \setminus \{0\}$. In particular, this group is linearly transitive if and only if $K(Q) = Q$, i.e. $Q$ is a skewfield. Similar descriptions apply to other groups of homologies and shears of $\mathcal{A}$.

DEFINITION 1.19. Let $Q$ be a right quasifield. The *distributor* and the *right* and the *middle nucleus* of $Q$ are defined by

$$D(Q) = \{t \in Q \mid x \circ (t + y) = x \circ t + x \circ y \text{ for all } x, y \in Q\}$$
$$N_r(Q) = \{c \in Q \mid (x \circ y) \circ c = x \circ (y \circ c) \text{ for all } x, y \in Q\}$$
$$N_m(Q) = \{c \in Q \mid (x \circ c) \circ y = x \circ (c \circ y) \text{ for all } x, y \in Q\}.$$

The quasifield $Q$ is called a *semifield* or *distributive quasifield* if $D(Q) = Q$. $Q$ is called a *nearfield* if $N_r(Q) = Q$ or equivalently $N_m(Q) = Q$.

Note that $Q$ is a semifield if and only if $Q$ satisfies both distributive laws and $Q$ is a nearfield if and only if the multiplication of $Q$ is associative. Semifields are also called division rings or division algebras since every semifield $Q$ is an algebra over its center $C(Q) = \{c \in K(Q) \mid c \circ x = x \circ c \text{ for all } x \in Q\}$.

One can also define the left nucleus of a quasifield. Recently, it was used by Kolb [68] to give an alternative proof of Corollary 1.23.

PROPOSITION 1.20. *Let $Q$ be a quasifield and let $\mathcal{A} = \mathcal{A}(Q)$ be the translation plane associated with $Q$. Put $S = \{0\} \times Q$ and denote the improper point of $S$ by $s$. Then*

$$\Sigma_{[s,S]} = \{\sigma_t : Q^2 \to Q^2 : (x, y) \mapsto (x, x \circ t + y) \mid t \in D(Q)\},$$

*where $\Sigma_{[s,S]}$ is the group of shears with axis $S$ and center $s$.*

*Proof.* Let $t \in D(Q)$ and let $x, m, b \in Q$. Then we compute

$$\sigma_t(x, x \circ m + b) = (x, x \circ t + x \circ m + b) = (x, x \circ (t + m) + b)$$

since $t \in D(Q)$. It follows that $\sigma_t$ maps lines of slope $m$ to lines of slope $t + m$ and hence is a collineation of $\mathcal{A}$. Obviously, $\sigma_t$ fixes all points on $S$ and all lines parallel to $S$. Thus, $\sigma_t$ is a shear with axis $S$ and center $s$.

Let now $\sigma : Q^2 \to Q^2$ be a shear with axis $S$ and center $s$. By Proposition 1.14 and Theorem 1.18, $\sigma$ is a semilinear mapping of the $K(Q)$-vector space $Q^2$. Since $\sigma$ fixes all points of $S$, it is even linear. As $\sigma$ fixes all lines of $\mathcal{A}$ parallel to $S$, it follows from elementary linear algebra that there exists a linear mapping $\kappa : Q \to Q$ such that

$$\sigma(x, y) = (x, \kappa(x) + y) \quad \text{for all } x, y \in Q.$$

For $m \in Q$ let $U_m = \{(x, x \circ m) \mid x \in Q\}$ denote the line with slope $m$ passing through the origin. Then

$$\sigma(U_m) = \{(x, \kappa(x) + x \circ m) \mid x \in Q\} = U_{\varphi(m)} = \{(x, x \circ \varphi(m)) \mid x \in Q\},$$

where $\varphi : Q \to Q$ is bijective. This implies

$$\kappa(x) + x \circ m = x \circ \varphi(m) \quad \text{for all } x, m \in Q.$$

For $m = 0$ this yields $\kappa(x) = x \circ t$, where $t = \varphi(0)$. Setting now $x = 1$ we get $t + m = \varphi(m)$. This implies

$$x \circ t + x \circ m = x \circ (t + m) \quad \text{for all } x, m \in Q$$

and hence $t \in D(Q)$. □

COROLLARY 1.21. *Let $A = A(Q)$ be the translation plane associated with the quasifield $Q$ and let $S = \{0\} \times Q$. Then $Q$ is a semifield if and only if the shear group $\Sigma_{[s,S]}$ is linearly transitive.*

A projective plane $\mathcal{P} = (P, \mathcal{L})$ is of *Lenz type* $V$ if there exists an incident point line pair $(p, L) \in P \times \mathcal{L}$ such that $\mathrm{LBF}(\mathcal{P}) = \{(q, L) \in P \times \mathcal{L} \,|\, q \in L\} \cup \{(p, M) \in P \times \mathcal{L} \,|\, p \in M\}$, where $\mathrm{LBF}(\mathcal{P})$ denotes the Lenz-Barlotti figure of $\mathcal{P}$. It follows from Corollary 1.21 that $\mathcal{P}$ is of Lenz type at least $V$ if and only if $\mathcal{P}$ is isomorphic to the plane over a suitable semifield.

PROPOSITION 1.22. *Let $Q$ be a quasifield and let $A = A(Q)$ be the translation plane associated with $Q$. Put $S = \{0\} \times Q$ and $W = U_0 = Q \times \{0\}$ and let $s$ and $w$ denote the improper points of $S$ and $W$, respectively. Then*

$$\Sigma_{[s,W]} = \{\vartheta_c : Q^2 \to Q^2 : (x,y) \mapsto (x, y \circ c) \,|\, c \in N_r(Q) \setminus \{0\}\}$$

*and*

$$\Sigma_{[w,S]} = \{\chi_c : Q^2 \to Q^2 : (x,y) \mapsto (x \circ c, y) \,|\, c \in N_m(Q) \setminus \{0\}\}.$$

*Proof.* Let $c \in N_r(Q) \setminus \{0\}$ and let $x, m, b \in Q$. Then we get

$$\vartheta_c(x, x \circ m + b) = (x, ((x \circ m) + b) \circ c) = (x, x \circ (m \circ c) + b \circ c)$$

since $c \in N_r(Q)$. It follows that $\vartheta_c$ maps the line of slope $m$ and $y$-intercept $b$ to the line of slope $m \circ c$ and $y$-intercept $b \circ c$. Moreover, $\vartheta$ fixes all points on the line $W$ and all lines parallel to $S$. It follows that $\vartheta$ is a collineation of $A$ with axis $W$ and center $s$.

Let now $\vartheta \in \Sigma_{[s,W]}$. Then $\vartheta$ is a semilinear mapping of the left $K(Q)$-vector space $Q^2$. Since $\vartheta$ fixes every point of $W$ and all lines parallel to $S$, it is even linear and there exists a linear mapping $\kappa : Q \to Q$ such that

$$\vartheta(x,y) = (x, \kappa(y)) \quad \text{for all } x, y \in Q.$$

For $m \in Q$ let $U_m = \{(x, x \circ m) \,|\, x \in Q\}$. Then there exists a bijection $\varphi : Q \to Q$ such that

$$\vartheta(U_m) = \{(x, \kappa(x \circ m)) \,|\, x \in Q\} = U_{\varphi(m)} = \{(x, x \circ \varphi(m)) \,|\, x \in Q\}.$$

It follows that

$$\kappa(x \circ m) = x \circ \varphi(m) \quad \text{for all } x, m \in Q.$$

Setting $x = 1$ we get $\kappa = \varphi$ and setting $m = 1$ we get $\kappa(x) = x \circ c$, where $c = \kappa(1)$.

Let $x, m \in Q$. Then we have

$$(x \circ m) \circ c = \kappa(x \circ m) = x \circ \kappa(m) = x \circ (m \circ c).$$

Thus $c \in N_r(Q)$.

We now show that $\Sigma_{[w,S]} = \{\chi_c \mid c \in N_m(Q) \setminus \{0\}\}$.

Let $c \in N_m(Q) \setminus \{0\}$ and let $c^{-1} \in Q$ denote the unique element with $c \circ c^{-1} = 1$. Let $x \in Q$, then we have $(x \circ c) \circ c^{-1} = x \circ (c \circ c^{-1}) = x \circ 1 = x$. It follows that $\varrho_{c^{-1}} \circ \varrho_c : Q \to Q$ is the identity mapping. Thus, $\varrho_c \circ \varrho_{c^{-1}}$ is also the identity mapping, i.e. there holds $(x \circ c^{-1}) \circ c = x$ for all $x \in Q$.

Let $x, m, b \in Q$. Then we have

$$\begin{aligned}
\chi_{c^{-1}}(x, x \circ m + b) &= (x \circ c^{-1}, x \circ m + b) \\
&= (x \circ c^{-1}, ((x \circ c^{-1}) \circ c) \circ m + b) \\
&= (x \circ c^{-1}, (x \circ c^{-1}) \circ (c \circ m) + b).
\end{aligned}$$

It follows that $\chi_{c^{-1}}$ maps the line of slope $m$ and $y$-intercept $b$ to the line of slope $c \circ m$ and $y$-intercept $b$. Thus, $\chi_{c^{-1}}$ is a collineation of $\mathcal{A}$ with axis $S$ and center $w$. Since $\chi_c = \chi_{c^{-1}}^{-1}$, we also have $\chi_c \in \Sigma_{[w,S]}$.

Let now $\chi \in \Sigma_{[w,S]}$. As above, there exists a linear mapping $\kappa : Q \to Q$ such that

$$\chi(x, y) = (\kappa(x), y) \quad \text{for all } x, y \in Q.$$

There exists a bijection $\varphi : Q \to Q$ such that

$$\begin{aligned}
\chi(U_m) &= \{(\kappa(x), x \circ m) \mid x \in Q\} = \{(x, \kappa^{-1}(x) \circ m) \mid x \in Q\} \\
&= U_{\varphi(m)} = \{(x, x \circ \varphi(m)) \mid x \in Q\}.
\end{aligned}$$

It follows that

$$\kappa^{-1}(x) \circ m = x \circ \varphi(m) \quad \text{for all } x, m \in Q.$$

We put $c = \kappa^{-1}(1)$ and by setting $m = 1$ and $x = 1$ we get $\kappa^{-1}(x) = x \circ c$ and $\varphi(m) = c \circ m$, respectively.

Let $x, m \in Q$. Then

$$(x \circ c) \circ m = \kappa^{-1}(x) \circ m = x \circ \varphi(m) = x \circ (c \circ m).$$

It follows that $c \in N_m(Q)$. Let $c^{-1} \in Q$ be the unique element with $c \circ c^{-1} = 1$. From the first part of the proof we infer that $c^{-1} \in N_m(Q)$ and $\kappa(x) = x \circ c^{-1}$ for all $x \in Q$. Thus $\chi = \chi_{c^{-1}}$. $\qquad \square$

COROLLARY 1.23. *Let $\mathcal{A} = \mathcal{A}(Q)$ be the translation plane associated with the quasifield $Q$. Put $S = \{0\} \times Q$ and $W = Q \times \{0\}$ and denote the improper points of $S$ and $W$ by $s$ and $w$, respectively. Then the following conditions are equivalent:*

(a) *The quasifield $Q$ is a nearfield.*

(b) *The homology group $\Sigma_{[s,W]}$ is linearly transitive.*

(c) *The homology group $\Sigma_{[w,S]}$ is linearly transitive.*

The following result is known as *Gingerich's exchange theorem.*

COROLLARY 1.24. *Let $A$ be a translation plane and let $S$ and $W$ be distinct lines of $A$ with improper points $s$ and $w$, respectively. Then the homology group $\Sigma_{[w,S]}$ is linearly transitive if and only if $\Sigma_{[s,W]}$ is linearly transitive.*

A coordinate free proof of Gingerich's theorem was obtained by Percsy [82].

A projective plane $\mathcal{P} = (P, \mathcal{L})$ is of *Lenz-Barlotti type IV.a2* if there exist two distinct points $p, q \in P$ such that $\mathrm{LBF}(\mathcal{P}) = \{(r, p \vee q) \mid r \in p \vee q\} \cup \{(p, M) \mid q \in M\} \cup \{(q, M) \mid p \in M\}$. Corollary 1.23 implies that the nearfield planes are precisely the planes of Lenz-Barlotti type at least IV.a2.

If one uses left quasifields instead of right quasifields, the right nucleus is replaced by the left nucleus and the distributor is defined by $\{t \in Q \mid (t+x) \circ y = t \circ y + x \circ y$ for all $x, y \in Q\}$. The middle nucleus remains the same. In the formulation of Proposition 1.20 and 1.22 the factors in all products that appear have to be exchanged.

## 1.4 Some Topological Tools

In this section we collect some results from topology which we shall need later on.

We start with

BROUWER'S THEOREM ON DOMAIN INVARIANCE. *Let $M$ and $N$ be topological manifolds of the same dimension.*

(a) *Let $U \subseteq M$ and $V \subseteq N$ be homeomorphic. Then $U$ is open in $M$ if and only if $V$ is open in $N$.*

(b) *Let $f : M \to N$ be continuous and bijective. Then $f$ is a homeomorphism.*

For a proof of this theorem, the reader is referred to the books by Dugundji [27: XVIII. 3], Eilenberg-Steenrod [29: XI. §3], Hocking-Young [60: 6.17] or Massey [79: III. §6].

Let $X$ and $X'$ be locally compact topological spaces and let $\tilde{X} = X \cup \{\infty\}$ and $\tilde{X}' = X' \cup \{\infty'\}$ denote their one point compactifications. A continuous mapping $f : X \to X'$ is called *proper* if $f$ admits a continuous extension $\tilde{f} : \tilde{X} \to \tilde{X}'$ with $\tilde{f}(\infty) = \infty'$. If $X$ and $X'$ are finite dimensional normed real vector spaces, then $f$ is proper if and only if the following holds: $\|f(x)\| \to \infty$ for $\|x\| \to \infty$.

vector spaces, then $f$ is proper if and only if the following holds: $\|f(x)\| \to \infty$ for $\|x\| \to \infty$.

PROPOSITION 1.25. *Let* $f : \mathbb{R}^n \to \mathbb{R}^n$ *be continuous and injective. Then* $f$ *is a homeomorphism if and only if* $f$ *is proper.*

*Proof.* This is a well-known result, but it is not easy to find a proof in the literature. So we supply one.

Since the one point compactification of a locally compact topological space is unique up to homeomorphism, every homeomorphism is proper.

Let now $f : \mathbb{R}^n \to \mathbb{R}^n$ be proper and injective. The one point compactification of $\mathbb{R}^n$ is homeomorphic to $\mathbb{S}_n$. The continuous extension $\tilde{f} : \mathbb{S}_n \to \mathbb{S}_n$ of $f$ is injective. Since $\mathbb{S}_n$ is compact, it follows that $\mathbb{S}_n$ and $\tilde{f}(\mathbb{S}_n)$ are homeomorphic. By Brouwer's theorem, $\tilde{f}(\mathbb{S}_n)$ is open and as a compact subset also closed in $\mathbb{S}_n$. Since $\mathbb{S}_n$ is connected, $\tilde{f}$ is surjective and hence a homeomorphism. It follows that $f$ is a homeomorphism as well. $\qquad\square$

## 1.5 Locally Compact Connected Translation Planes

A *topological projective plane* is a projective plane whose point and line set are equipped with topologies such that the geometric operations of joining points and intersecting lines are continuous on their domains of definition. A *topological affine plane* is defined analogously with the additional requirement that construction of parallels is also continuous. Fundamental information on topological planes can be obtained from Salzmann [87], Grundhöfer and Salzmann [37] or the books by Prieß-Crampe [85: 5. §2] or Salzmann et al. [91].

The topologies of the point and the line space of a topological projective or affine plane are uniquely determined by the neighbourhood filter of one point in the topology induced on one line through that point. In particular, the topology of the point space determines the topology of the line space and vice versa.

If the point space of a topological projective or affine plane is locally compact and connected, its line space enjoys the same properties.

The kernel of a locally compact connected translation plane is a locally compact connected skewfield and hence isomorphic to $\mathbb{R}$, $\mathbb{C}$ or $\mathbb{H}$ by Pontrjagin's theorem. Each of these skewfields is a finite-dimensional real vector space. Since a locally compact real vector space is of finite dimension, the point space of a locally compact connected affine translation plane is a finite-dimensional real vector space.

Let $V$ be an $n$-dimensional real vector space. The set $G_k(V)$ of all $k$-dimensional subspaces of $V, 0 < k < n$, is a topological manifold as homogeneous space of the Lie group $GL(V)$. It is called a *Grassmann manifold*.

The following result was implicitly proved by Breuning [16]. For a short and simple proof the reader is referred to Löwen [71].

PROPOSITION 1.26. *Let $V$ be a real vector space of finite dimension $2n$ and let $\mathcal{B}$ be a spread of $V$. Then the following conditions are equivalent:*

(a) *The affine translation plane associated with $\mathcal{B}$ is a topological affine plane.*

(b) *The projective translation plane associated with $\mathcal{B}$ is a topological projective plane.*

(c) *$\mathcal{B}$ is a compact subset of the Grassmann manifold $G_n(V)$.*

A spread with these properties is called a *topological spread*.

Let $V$ be a $2n$-dimensional vector space over a finite field $GF(q)$ and let $\mathcal{B}$ be a partial spread of $V$. Then a simple counting argument shows that $\mathcal{B}$ is a spread if and only if $|\mathcal{B}| = q^n + 1$. A similar result holds for topological spreads of real vector spaces.

PROPOSITION 1.27. *Let $\mathcal{B}$ be a partial spread of a $2n$-dimensional real vector space $V$. Then $\mathcal{B}$ is a topological spread of $V$ if and only if $\mathcal{B}$ is homeomorphic to $\mathbb{S}_n$ as a subspace of the Grassmann manifold $G_n(V)$.*

*Proof.* Let $\mathcal{B}$ be a topological spread of the $2n$-dimensional real vector space $V$. Then $\mathcal{B}$ can be considered as a line pencil of the translation plane associated with $\mathcal{B}$. Hence $\mathcal{B}$ is homeomorphic to $\mathbb{S}_n$, cf. e.g. Salzmann [87: 7.23].

To prove the converse, we show first that the set $\mathcal{B}_1$ of all 1-dimensional subspaces of $V$ which are contained in an element of $\mathcal{B}$ is compact. By [71: Proposition] the set $I_{1,n} = \{(U_1, U_2) \in G_1(V) \times G_n(V) \,|\, U_1 \leq U_2\}$ is compact. The set $\mathcal{B}_1$ is obtained from $I_{1,n} \cap G_1(V) \times \mathcal{B}$ by projection onto the first factor. Thus $\mathcal{B}_1$ is compact as well.

Let now $S \in \mathcal{B}$ and $x \in V \setminus S$. Consider the mapping

$$\varphi : \mathcal{B} \setminus \{S\} \to S + x : U \mapsto U \cap (S + x).$$

Since the elements of $\mathcal{B}$ are mutually complementary, $\varphi$ is injective. By assumption $\mathcal{B} \setminus \{S\}$ is homeomorphic to $\mathbb{R}^n$. So $\varphi$ is a continuous injective mapping between two topological spaces which are both homeomorphic to $\mathbb{R}^n$. By Brouwer's theorem $\varphi(\mathcal{B} \setminus \{S\})$ is open in $S + x$. If we can show that $\varphi(\mathcal{B} \setminus \{S\})$ is also closed in $S + x$, then $\varphi$ is bijective and consequently $x$ is contained in an element of $\mathcal{B} \setminus \{S\}$. So let $(y_k)$ be a sequence in $\varphi(\mathcal{B} \setminus \{S\})$ that converges to a vector $y \in S + x$. Since $\mathcal{B}_1$ is compact, there exists $W \in \mathcal{B}$ such that $(y) \subseteq W$. From $x \in V \setminus S$ we infer that $W \neq S$. It follows that $y \in \varphi(\mathcal{B} \setminus \{S\})$ and hence $\varphi(\mathcal{B} \setminus \{S\})$ is closed in $S + x$. $\square$

For 4-dimensional vector spaces this result was already proved by Gluck and Warner [34: 6]. Besides Brouwer's theorem their proof also uses the action of the orthogonal group on the Grassmann manifold. A generalization of Proposition 1.27 to systems of mutually skew $k$-dimensional subspaces of an $n$-dimensional real vector space, $n \neq 2k$, was obtained by Löwen [72].

Using transfinite induction, Bernardi [6] has shown that each even dimensional vector space over an infinite field contains a spread. However, for the existence of topological spreads there are strong restrictions. The following result is due to Salzmann [87: 7.23].

PROPOSITION 1.28. *Let V be a real vector space of dimension 2n which contains a topological spread. Then we have* $n \in \{1, 2, 4, 8\}$.

For locally compact connected translation planes whose kernel is strictly larger than $\mathbb{R}$ the possible dimensions are further reduced by the following result of Buchanan and Hähl [22].

PROPOSITION 1.29. *Let V be a complex vector space of (complex) dimension 2n which contains a topological spread. Then we have* $n = 1$ *or* $n = 2$.
   *Let V be a quaternionic vector space of (quaternionic) dimension 2n which contains a topological spread. Then we have* $n = 1$.

This proposition says essentially that the kernel of a locally compact 16-dimensional translation plane is isomorphic to $\mathbb{R}$.

In particular, the kernel of a locally compact connected translation plane is commutative, with the quaternion plane being the only exception.

## 1.6 The Transposition of Translation Planes

Let $V$ be a vector space over a skewfield $F$. Associated with $V$ is the dual space $V^*$ which consists of all linear functionals $\varphi : V \to F$. The *annulator* of a subspace $U \leq V$ is defined by $U^\circ = \{\varphi \in V^* \,|\, \varphi(x) = 0 \text{ for all } x \in U\}$.

LEMMA 1.30. *Let $\mathcal{B}$ be a spread of $V$. Then $\mathcal{B}^\circ = \{U^\circ \,|\, U \in \mathcal{B}\}$ is a partial spread of $V^*$.*

*Proof.* For all $U_1, U_2 \leq V$ we have

$$(U_1 + U_2)^\circ = U_1^\circ \cap U_2^\circ \quad \text{and} \quad (U_1 \cap U_2)^\circ = U_1^\circ + U_2^\circ.$$

It follows that any two elements of $\mathcal{B}^\circ$ are complementary and hence $\mathcal{B}^\circ$ is a partial spread. $\square$

If $\mathcal{B}^\circ$ is even a spread we say that $\mathcal{B}$ is a *dual spread*. Every even dimensional vector space $V$ over an infinite skewfield $F$ admits a spread which is not a dual spread. For countable $F$ this was proved by Bruen and Fisher [21] and the general case was handled by Bernardi [6].

For finite fields we have the following

PROPOSITION 1.31. *Let $V$ be a finite dimensional vector space over a finite field $F$ and let $\mathcal{B}$ be a spread of $V$. Then $\mathcal{B}$ is also a dual spread.*

*Proof.* Since $\mathcal{B}^\circ$ is a partial spread and $|\mathcal{B}^\circ| = |\mathcal{B}|$, this follows from an easy counting argument. $\qquad\square$

An analogous result for topological spreads of real vector spaces was proved by Buchanan and Hähl [23].

**PROPOSITION 1.32.** *Let $V$ be a finite dimensional real vector space and let $\mathcal{B}$ be a topological spread of $V$. Then $\mathcal{B}$ is also a dual spread.*

*Proof.* $\mathcal{B}^\circ$ is a partial spread of $V^*$ and the mapping

$$^\circ : G_n(V) \to G_n(V^*) : U \mapsto U^\circ$$

is a homeomorphism. Hence, $\mathcal{B}^\circ$ is homeomorphic to $\mathcal{B}$ and then $\mathcal{B}^\circ$ is a spread by Proposition 1.27. $\qquad\square$

If $\mathcal{B}^\circ$ is a spread of $V^*$ then the translation plane associated with $\mathcal{B}^\circ$ is called the *transposed translation plane* of the plane associated with $\mathcal{B}$. The explanation for this name is given by the following

**PROPOSITION 1.33.** *Let $F$ be a commutative field and let $V$ be a vector space over $F$ with $dim(V) = 2n < \infty$. Let $\mathcal{B}$ be spread of $V$ which is also a dual spread. If $\mathcal{M}$ is a spread set for $\mathcal{B}$ then $\mathcal{M}^\circ = \{A^t \mid A \in \mathcal{M}\}$ is a spread set for $\mathcal{B}^\circ$.*

*Proof.* Let $V$ be identified with $F^n \times F^n$ such that $S = \{0\} \times F^n \in \mathcal{B}$ and $\mathcal{B} \setminus \{S\} = \{\{(x, Ax) \mid x \in F^n\} \mid A \in \mathcal{M}\}$. We define a non-degenerate alternating bilinear form $B : V \times V \to F$ by

$$B((x_1, y_1), (x_2, y_2)) = x_1^t \cdot y_2 - y_1^t \cdot x_2.$$

For $U \leq V$ we denote by $U^\perp = \{v \in V \mid B(u, v) = 0 \text{ for all } u \in U\}$ the orthogonal space of $U$. Since $B$ induces an isomorphism between $V$ and $V^*$, the set $\mathcal{B}^* = \{U^\perp \mid U \in \mathcal{B}\}$ is a spread of $V$ which is equivalent to $\mathcal{B}^\circ$.

For $A \in M_{n,n}(F)$ we have

$$B((x, Ax), (y, A^t y)) = x^t \cdot A^t y - (Ax)^t \cdot y = 0 \quad \text{for all } x, y \in F^n.$$

It follows that

$$\{(x, Ax) \mid x \in F^n\}^\perp = \{(y, A^t y) \mid y \in F^n\}$$

since both spaces are $n$-dimensional. Moreover we have $S^\perp = S$. Hence $\mathcal{M}^\circ = \{A^t \mid A \in \mathcal{M}\}$ is a spread set for $\mathcal{B}^\circ$. $\qquad\square$

# 2. Spreads of 3-dimensional Projective Spaces

In this chapter $F$ always denotes a skewfield, $\mathcal{P} = (P, \mathcal{L})$ a 3-dimensional projective space over $F$ and $\mathcal{B}$ a spread of $\mathcal{P}$.

## 2.1 The Klein Quadric

In this section we assume $F$ to be commutative. Then we can describe the lines of $\mathcal{P}$ using Plücker coordinates and the Klein quadric.

Our approach follows the books of Blaschke [14: 59] and Pickert [83: 34], cf. also Segre [96: Chap. 15] and Burau [24: VIII].

Let $V$ denote the 4-dimensional vector space underlying $\mathcal{P}$ and let $N$ be the 6-dimensional vector space of alternating bilinear forms on $V$. Note that the radical of an alternating bilinear form on $V$ has even dimension.

Let $L$ be a 2-dimensional subspace of $V$, and let $x, y \in V$ be two vectors which together with $L$ generate $V$. Now let $B, C \in N$ be two forms such that $L$ is contained in the radical of both of them. We put $b = B(x, y)$ and $c = C(x, y)$. A simple calculation shows that $bC - cB$ is the zero form. So we can define a mapping from the set of all lines of $\mathcal{P}$ to the point set of the projective space over $N$ by sending a line $L$ to the set of all alternating bilinear forms on $V$ that contain $L$ in their radical. It is called the *Plücker mapping*.

After choosing a basis of $V$ the alternating bilinear forms are represented by skew symmetric matrices having zeros on the main diagonal. The determinant of such a matrix is the square of a quadratic form in the coefficients, the so called *Pfaffian*. For a proof of this fact we refer to Jacobson [62: Theorem 6.4]. It is also easy to compute directly that

$$\det \begin{pmatrix} 0 & x_1 & x_2 & x_3 \\ -x_1 & 0 & x_4 & x_5 \\ -x_2 & -x_4 & 0 & x_6 \\ -x_3 & -x_5 & -x_6 & 0 \end{pmatrix} = (x_1 x_6 - x_2 x_5 + x_3 x_4)^2.$$

The singular alternating bilinear forms are represented by matrices having determinant equal to zero. So the image of the Plücker mapping is a quadric, called the *Klein quadric*. The set of all maximal isotropic subspaces on the Klein quadric decomposes into two families. Two different maximal isotropic subspaces belong to the same family if and only if their intersection is 1-dimensional. These two families of subspaces can be identified with the points and planes of $\mathcal{P}$ in such a way that the incidence of points and planes with lines is preserved. Moreover, a point and a plane of $\mathcal{P}$ are incident if and only if their corresponding subspaces on the Klein quadric have a 2-dimensional intersection. Two lines of $\mathcal{P}$ intersect if and only if their images under the Plücker mapping are orthogonal with respect to the bilinear form associated with the Pfaffian.

So a spread of $\mathcal{P}$ corresponds under the Plücker mapping to a set of points on the Klein quadric that has exactly one point in common with every maximal isotropic subspace in one of the two families.

A subset of a quadric is called an *ovoid* if it has exactly one point in common with each maximal isotropic subspace of the quadric. Using this notion, we can say that the ovoids of the Klein quadric are precisely the images under the Plücker mapping of the spreads of $\mathcal{P}$ which are also dual spreads.

The Plücker mapping and the Klein quadric are defined geometrically and they are uniquely determined by their properties, cf. e.g. [100: Theorem 8.6] or [51: Satz 4.1]. Their actual appearance depends on the description of the vector space $V$ and the 2-dimensional subspaces of $V$.

Assume for example that $V = X \times X$ where $X$ is a 2-dimensional vector space. Then each 2-dimensional subspace of $V$ whose intersection with $S = \{0\} \times X$ is trivial is the graph of a linear mapping $\varphi : X \to X$. We can describe the Klein quadric using the vector space $\mathrm{End}_F(X) \times F^2$. The Pfaffian is given by $\mathrm{Pf}(\lambda, x, y) = \det(\lambda) - xy$ for $\lambda \in \mathrm{End}_F(X)$ and $x, y \in X$. The graph of $\lambda \in \mathrm{End}_F(X)$ gets Plücker coordinates $\langle (\lambda, 1, \det(\lambda)) \rangle$ and the subspace $S$ has Plücker coordinates $\langle (0, 0, 1) \rangle$.

A generalization of the Klein quadric to projective spaces over proper skewfields was developed by Havlicek [53].

## 2.2 Transversal Mappings

In this section we generalize a method of Ostrom [81] for the description of spreads from finite projective spaces to arbitrary projective spaces. Unlike Ostrom, who argues algebraically, we make a direct geometric approach.

Let $S \in \mathcal{B}$ be a fixed line and choose two planes $E_1$ and $E_2$ of $\mathcal{P}$ whose intersection is $S$. Furthermore, let $p$ be a point of $\mathcal{P}$ that is contained neither in $E_1$ nor in $E_2$. Let $\pi : E_2 \to E_1$ denote the projection with center $p$. The affine plane we get from $E_i$ by deleting the line $S$ is denoted by $E_i'$ for $i = 1, 2$. Since $\mathcal{B}$ is a spread, each line $G \in \mathcal{B} \setminus \{S\}$ intersects the planes $E_1'$ and $E_2'$ in exactly one point $a(G) = G \cap E_1$ and $b(G) = G \cap E_2$, respectively. By setting $f(a(G)) = \pi(b(G))$ we get a bijection $f : E_1' \to E_1'$.

Let $G_1, G_2 \in \mathcal{B} \setminus \{S\}$ be two different lines. Since $G_1$ and $G_2$ do not intersect, the lines $H_1 = a(G_1) \vee a(G_2)$ and $H_2 = b(G_1) \vee b(G_2)$ are also disjoint by the Veblen-Young axiom. Hence the intersection point of the lines $H_1$ and

$\pi(H_2)$ is not contained in $S$. So $f$ has the property that for any two different points $a_1, a_2 \in E_1'$ we have $a_1 \vee a_2 \not\parallel f(a_1) \vee f(a_2)$.

Next we want to find out what it means for $f$ that every point of $\mathcal{P}$ is contained in an element of $\mathcal{B}$. Let $q$ be a point of $\mathcal{P}$ which is not contained in the union of $E_1$ and $E_2$ and let $\pi_q : E_2 \to E_1$ denote the projection with center $q$. Obviously, the point $q$ is contained in the line $G \in \mathcal{B} \setminus \{S\}$ if and only if $\pi_q(b(G)) = a(G)$. By definition we have $b(G) = \pi^{-1}(f(a(G)))$. It follows that for each point $q$ which is not contained in the union of $E_1$ and $E_2$ there exists a line $G \in \mathcal{B} \setminus \{S\}$ such that $a(G) = \pi_q(\pi^{-1}(f(a(G))))$. Note that the mappings $\pi_q \circ \pi^{-1}$ are precisely the dilatations of the affine plane $E_1'$.

The properties just stated motivate the following

DEFINITION 2.1. Let $E$ be an affine plane over a skewfield $F$. A bijective mapping $f : E \to E$ is called *transversal* if it has the following properties:

(T1) For any two different points $a_1, a_2 \in E$ the lines $a_1 \vee a_2$ und $f(a_1) \vee f(a_2)$ are not parallel.

(T2) For each dilatation $\delta$ of the affine plane $E$ the mapping $\delta \circ f$ has a fixed point.

Using this concept we can say that each spread of $\mathcal{P}$ can be described by a transversal mapping.

There is also a reverse construction. Let $E$ be a desarguesian affine plane and let $f : E \to E$ be a transversal mapping. We embed $E$ into a 3-dimensional projective space $\mathcal{P}$ and denote the improper line of the projective extension of $E$ by $S$. Furthermore, we choose a plane $E_2$ of $\mathcal{P}$ which contains $S$ and a point $p$ which is not contained in the union of $E$ and $E_2$. Let $\pi : E_2 \to E$ denote the projection with centre $p$. Then $\mathcal{B} = \{S\} \cup \{a \vee \pi^{-1}(f(a)) \mid a \in E\}$ is a spread of $\mathcal{P}$.

Summarizing we get the following

THEOREM 2.2. *Every spread of a 3-dimensional projective space over a skewfield $F$ can be described by a transversal mapping of the affine plane over $F$. Conversely, every transversal mapping of an affine plane over a skewfield $F$ defines a spread of the 3-dimensional projective space over $F$.*

For finite fields this proposition was proved by Ostrom [81]. In this case a simple counting argument shows that (T1) implies (T2) and hence Ostrom requires only (T1).

In order to construct a translation plane from a transversal mapping, one has to choose two different planes and a point which is not contained in their union from a 3-dimensional projective space. The collineation group of the projective space acts transitively on the set of these configurations and hence each transversal mapping determines a unique translation plane.

Our next aim is to get a coordinate description of the spread which is associated with a transversal mapping. Let $V$ be a 4-dimensional right vector space over $F$ and denote the corresponding projective space by $\mathcal{P}$. After specifying a basis of $V$ we can think of the elements of $V$ as column vectors. We choose the basis such that the following hold:

$$S = \{(0, 0, u, v)^t \mid u, v \in F\},$$
$$E_1 = \{(x, 0, u, v)^t \mid x, u, v \in F\},$$
$$E_2 = \{(0, y, u, v)^t \mid y, u, v \in F\},$$
$$p = \langle (1, -1, 0, 0)^t \rangle.$$

For the projection $\pi : E_2 \to E_1$ we get

$$\pi(\langle (0, y, u, v)^t \rangle) = \langle (y, 0, u, v)^t \rangle.$$

We now choose coordinates in the affine plane $E_1'$ such that the point $(a, b) \in F^2$ corresponds to $\langle (1, 0, a, b)^t \rangle$. Let $f : E_1' \to E_1'$ be a transversal mapping. The spread $\mathcal{B}$ associated with $f$ contains $S$ and the subspaces

$$L(a, b) = \langle (1, 0, a, b)^t, (0, 1, f_1(a, b), f_2(a, b))^t \rangle$$
$$= \{(x, y, ax + f_1(a, b)y, bx + f_2(a, b)y)^t \mid x, y \in F\}, \quad a, b \in F.$$

$L(a, b)$ is the graph of the linear mapping

$$F^2 \to F^2 : \begin{pmatrix} x \\ y \end{pmatrix} \mapsto \begin{pmatrix} a & f_1(a, b) \\ b & f_2(a, b) \end{pmatrix} \begin{pmatrix} x \\ y \end{pmatrix}.$$

Ostrom first constructs the dual of the translation plane associated with a transversal mapping $f$ as follows: Points are the elements of $F^4$, lines are the sets

$$\{(x_1, x_2, y_1, y_2) \mid y_1 = x_1\alpha + b_1, y_2 = x_2\alpha + b_2\}, \alpha, b_1, b_2 \in F, \quad \text{and} \quad S,$$

as well as the sets

$$\{(x_1, x_2, y_1, y_2) \mid y_1 = f_1(x_1, x_2)\alpha + x_1\beta + b_1, y_2 = f_2(x_1, x_2)\alpha + x_2\beta + b_2\}$$

for $\alpha, \beta, b_1, b_2 \in F$. We may think of the vectors $(\alpha, \beta, b_1, b_2)^t \in F^4$ as points of the translation plane. Using this description the spread contains $S$ and the graphs of the linear mappings

$$F^2 \to F^2 : \begin{pmatrix} \alpha \\ \beta \end{pmatrix} \mapsto \begin{pmatrix} -f_1(x_1, x_2) & -x_1 \\ -f_2(x_1, x_2) & -x_2 \end{pmatrix} \begin{pmatrix} \alpha \\ \beta \end{pmatrix}$$

with $x_1, x_2 \in F$. The linear mapping from $F^4$ to itself given by $(\alpha, \beta, b_1, b_2)^t \mapsto (-\beta, -\alpha, b_1, b_2)^t$ maps Ostrom's matrices to the matrices

$$\begin{pmatrix} x_1 & f_1(x_1, x_2) \\ x_2 & f_2(x_1, x_2) \end{pmatrix}, x_1, x_2 \in F.$$

These considerations show that our coordinate free approach to transversal mappings yields the same results as Ostrom's algebraic approach.


## 2.3 Betten's Variant

In [8] Betten defines transversal homeomorphisms of the real affine plane and uses them to describe locally compact 4-dimensional translation planes. In his definition of transversality (T2) is replaced by the condition that for any two non-parallel lines $H_1, H_2$ of $E$ we have $|H_1 \cap f(H_2)| = 1$. In general, this property is not equivalent to (T2). In this section we investigate Betten's condition geometrically in arbitrary projective spaces. We call mappings satisfying Betten's condition *-transversal in order to distinguish them from the transversal mappings.

It turns out that Betten's condition actually describes dual spreads instead of spreads. For this reason we replace the projective space $\mathcal{P}$ by its dual $\mathcal{P}^*$. Then $\mathcal{B}$ becomes a dual spread, i.e. a set $\mathcal{B}^*$ of lines of $\mathcal{P}^*$ having the property that each plane of $\mathcal{P}^*$ contains exactly one element of $\mathcal{B}^*$.

Just like in the last section we choose two different planes $E_1$ and $E_2$ of $\mathcal{P}^*$, which intersect in a line $S^* \in \mathcal{B}^*$. Furthermore, we select a point $p$ which is not contained in the union of $E_1$ and $E_2$. The mapping $\pi$, the points $a(G)$ and $b(G)$ for $G \in \mathcal{B}^* \setminus \{S^*\}$, and the mapping $f$ are defined as in the last section. It should be noted that the domain $D$ where $f$ is defined can be a proper subset of $E_1'$ since not every point of $E_1'$ must be contained in an element of $\mathcal{B}^*$.

As two different lines in $\mathcal{B}^*$ do not intersect, $f$ satisfies the condition (T1), but only for $a_1, a_2 \in D$.

Next we want to find out what it means for $f$ that each plane of $\mathcal{P}^*$ contains a line of $\mathcal{B}^*$. To this end we look at a line $H$ of $E_1'$ and the pencil of planes through $H$. Every plane $E$ of this pencil with $E \neq E_1$ intersects $E_2$ in a line. The image of this line under $\pi$ is a line in $E_1$, which is parallel to $H$. So $f(H \cap D)$ intersects each line which is parallel to $H$ non-trivially. In particular, we have $H \cap D \neq \emptyset$.

These considerations motivate the following

DEFINITION 2.3. Let $E$ be a desarguesian affine plane and let $D$ be a subset of $E$ which intersects each line of $E$ non-trivially. For a line $H$ of $E$ set $H' = H \cap D$. A mapping $f : D \rightarrow E$ is called *-transversal if $f$ has the following properties:

(T1*) For any two different points $a_1, a_2 \in D$ the lines $a_1 \vee a_2$ and $f(a_1) \vee f(a_2)$ are not parallel.

(T2*) For any two parallel lines $H_1, H_2$ of $E$ we have $H_1 \cap f(H_2') \neq \emptyset$.

Theorem 2.2 remains true if we replace the term transversal by *-transversal.

In [8] Betten studies topological spreads of the 3-dimensional real projective space using *-transversal mappings of the real affine plane. He shows that in this case the set $D$ is open and convex in $\mathbb{R}^2$. Since $D$ also intersects each line of $\mathbb{R}^2$ non-trivially, he gets $D = \mathbb{R}^2$. This result implies that every topological spread of a 3-dimensional real projective space is also a dual spread. The analogue of this theorem for topological spreads of finite-dimensional real projective spaces was proved by Buchanan and Hähl [23], cf. Proposition 1.32.

In general $D$ is a proper subset of $E$. Bruen and Fisher [21] have proved that every 3-dimensional projective space over a countable infinite field contains a spread which is not a dual spread. This result was generalized by Bernardi [6] to projective spaces of odd dimension over infinite fields of arbitrary cardinality. Consider now a spread of a 3-dimensional dual projective space $\mathcal{P}^*$ which is not a spread of $\mathcal{P}$. Then there is a point $q$ of $\mathcal{P}$ which is not on any line of the spread. If we choose the plane $E_1$ on which the $*$-transversal mapping is defined in such a way that $q \in E_1$ then $D \neq E_1'$.

Next we want to give a coordinate description of the spread associated with a $*$-transversal mapping. Let $V$ be a 4-dimensional right vector space over $F$ and denote its dual by $V^*$. The projective space asscociated with $V$ is denoted by $\mathcal{P}$. Then we can identify $\mathcal{P}^*$ with the projective space over $V^*$. Select dual bases of $V$ and $V^*$. The elements of $V$ and $V^*$ then correspond to column and row vectors, respectively. Furthermore, the image of a vector $(x, y, u, v)^t \in V$ under a linear functional $(a, b, c, d) \in V^*$ is given by $ax + by + cu + dv$. Assume that the basis of $V^*$ is chosen such that

$$S^* = \{(a, b, 0, 0) \mid a, b \in F\},$$
$$E_1 = \{(a, b, c, 0) \mid a, b, c \in F\},$$
$$E_2 = \{(a, b, 0, d) \mid a, b, d \in F\},$$
$$p = \langle (0, 0, 1, -1) \rangle.$$

The mapping $\pi : E_2 \to E_1$ is given by

$$\pi(\langle (a, b, 0, d) \rangle) = \langle (a, b, d, 0) \rangle.$$

In order to get coordinates in the affine plane $E_1'$ we set $c = -1$, i.e. we identify the points $(a, b) \in F^2$ and $\langle (a, b, -1, 0) \rangle$. Let now $f : D \to E_1'$ be a $*$-transversal mapping. Then the spread $\mathcal{B}^*$ associated with $f$ consists of $S^*$ and the lines

$$L(a, b)^* = \langle (a, b, -1, 0), (f_1(a, b), f_2(a, b), 0, -1) \rangle, (a, b) \in D.$$

We have

$$L(a, b) = L(a, b)^{*\circ} = \{(x, y, u, v)^t \in V \mid u = ax + by, v = f_1(a, b)x + f_2(a, b)y\}$$

and $\quad S = S^{*\circ} = \{(0, 0, u, v)^t \mid u, v \in F\}$.

Thus we get the following description of the spread associated with a $*$-transversal mapping. One element of the spread ist $S$, and the other elements are the graphs of the linear mappings

$$F^2 \to F^2 : \begin{pmatrix} x \\ y \end{pmatrix} \mapsto \begin{pmatrix} a & b \\ f_1(a,b) & f_2(a,b) \end{pmatrix} \begin{pmatrix} x \\ y \end{pmatrix}$$

with $(a,b) \in D$.

It should be noted that Betten gets exactly the transposed matrices of the matrices given above. This is due to the fact that he actually studies spreads of the dual space.

For the special case that $f$ is a collineation the results of this section were also obtained by Havlicek [52].

## 2.4 Indicator Sets

In this section we describe some ideas which are due to Bruen [20].

Let $F$ be a skewfield and let $L$ be an extension skewfield of $F$ which has rank 2 as a right vector space over $F$. It might well happen that the rank of $L$ as a left vector space over $F$ is different from 2; cf. Cohn [25: 5.6.1] for examples of this phenomenon. Bruen's idea now is as follows. Identify the affine plane over $F$ with $L$ and view the graph of a transversal mapping as a subset of the affine plane over $L$. Bruen calls this graph an *indicator set* since, in a certain sense, it indicates a spread.

Let $f : L \to L$ be a transversal mapping. The graph of $f$ is denoted by $\mathcal{J}$.

Let us examine what the conditions (T1) and (T2) look like if we formulate them using indicator sets. Assume that the affine plane over $L$ is defined over a 2-dimensional left vector space. Then the lines of this affine plane are the sets

$$\{(c, w) \mid w \in L\}, c \in L \quad \text{(lines with slope } \infty) \quad \text{and}$$

$$\{(z, zm + b) \mid z \in L\}, m, b \in L \quad \text{(lines with slope } m).$$

Since our coordinates are taken from a left vector space, the slope of a line naturally appears on the right.

Consider now the line which connects two different points $(x, f(x))$ and $(y, f(y))$ in the affine plane over $L$. We get this line by solving the simultaneous equations

$$f(x) = xm + b \quad \text{and} \quad f(y) = ym + b$$

for $m$. Subtraction of these equations yields

$$f(x) - f(y) = (x - y)m.$$

Since the lines $x \vee y$ and $f(x) \vee f(y)$ are not parallel in the affine plane $L$ over $F$, the vectors $f(x) - f(y)$ and $x - y$ are linearly independent in the right $F$-vector space $L$. Hence the last equation has no solution $m \in F$. So the slope of the line $(x, f(x)) \vee (y, f(y))$ in the affine plane over $L$ is not contained in $F$.

The dilatations of the affine plane $L$ over $F$ are precisely the mappings $\delta : L \to L : x \mapsto (x - t)s^{-1}$ where $s \in F^{\times}$ and $t \in L$. Condition (T2) states that

$\delta \circ f$ has at least one fixed point. Obviously, this condition is satisfied if and only if the equation $f(x) = xs + t$ has at least one solution $x \in L$. It follows that each line of the affine plane over $L$ whose slope is contained $F^{\times}$ intersects $\mathcal{J}$ in at least one point. Since $f$ is bijective, each line of slope 0 or $\infty$ also intersects $\mathcal{J}$ in at least one point.

These considerations lead us to the following

DEFINITION 2.4. Let $F$ be a skewfield and let $L$ be an extension skewfield of $F$ whose rank as a right vector space over $F$ is 2. Assume that the affine plane over $L$ is coordinatized by a 2-dimensional left vector space. Then a subset $\mathcal{J}$ of the affine plane over $L$ is called an *indicator set* if the following conditions are satisfied:

(I1) For any two distinct points $z, w \in \mathcal{J}$ the slope of the line $z \vee w$ is not contained in $F \cup \{\infty\}$.

(I2) Each line of the affine plane over $L$ whose slope is contained in $F \cup \{\infty\}$ intersects $\mathcal{J}$ in at least one point.

Since (I1) and (I2) are equivalent to (T1) and (T2), the analogue of Theorem 2.2 also holds for indicator sets.

Bruen only treats finite fields. He defines indicator sets using (I1) and requiring $|\mathcal{J}| = |L|$. It is easy to see that in the finite case these two conditions imply (I2).

Bruen's construction of the spread from the indicator set is slightly different from ours. He embeds the projective space $\mathcal{P}$ over $F$ in the projective space $\hat{\mathcal{P}}$ over $L$. Furthermore, he chooses a plane $E$ of $\hat{\mathcal{P}}$ which has a line $S$ in common with $\mathcal{P}$. Let $E'$ be the affine plane which is obtained from $E$ by deleting $S$. Assume now that $\mathcal{J}$ is an indicator set in the affine plane $E'$. We can view the slope of a line $H$ of $E'$ as the intersection point of $H$ and $S$. Moreover, we can identify $F \cup \{\infty\}$ with the set of the points of $S$ which belong to $\mathcal{P}$.

Each point of $\hat{\mathcal{P}}$ which does not belong to $\mathcal{P}$ is on exactly one line of $\mathcal{P}$. The set $\mathcal{B}$ consists of $S$ and of all lines of $\mathcal{P}$ which have a point in common with $\mathcal{J}$. Let $D$ be a plane of $\mathcal{P}$ which does not contain $S$. Then we can think of $E \cap D$ as a line of $E'$ whose slope is in $F \cup \{\infty\}$. Because of (I2) we have $|D \cap E \cap \mathcal{J}| \geq 1$ and by construction $D$ contains exactly one element of $\mathcal{B}$. Hence $\mathcal{B}$ is a spread of $\mathcal{P}^*$. Since Bruen only looks at projective spaces over finite fields, $\mathcal{B}$ also is a spread of $\mathcal{P}$. The corresponding translation plane is the transposed of the plane which was constructed from $\mathcal{J}$ by the method of Ostrom. This fact was also realized by Bruen [20: 5.A].

Sherk [97] and Lunardon [76] have generalized the notion of indicator set in order to describe also spreads of higher dimensional projective spaces.

## 2.5  $L_1$-indicator Sets

For the definition of an indicator set we need the projective line over $L$ and the projective line over the subfield $F$. Consider now a collineation $\chi$ of the affine plane over $L$ and its effect on an indicator set $\mathcal{J}$. Let $\hat{\chi}$ denote the mapping induced by $\chi$ on the set of slopes $L \cup \{\infty\}$. The set $\chi(\mathcal{J})$ has properties which are similar to those of an indicator set. One only has to replace the set $F \cup \{\infty\}$ by $\hat{\chi}(F \cup \{\infty\})$ in the formulation of the properties (I1) and (I2). We call $\chi(\mathcal{J})$ a $\hat{\chi}(F \cup \{\infty\})$-indicator set.

From now on we assume that $F$ and $L$ are commutative and that $L$ is separable over $F$. Let $^-$ denote the corresponding involutorial automorphism of $L$. The *norm* of an element $z \in L$ is defined by $N(z) = z\bar{z}$. With the pair of fields $(F,L)$ we can associate an *inversive plane* $\Sigma(F, L)$, cf. e.g. Benz [5: III. §2]. The point set of this inversive plane is $L \cup \{\infty\}$ and the circles are the images of $F \cup \{\infty\}$ under the elements of the group $\mathrm{PGL}_2(L)$. Apart from $F \cup \{\infty\}$ there is another distinguished circle, namely the *unit circle*. It is defined by $L_1 = \{z \in L \mid N(z) = 1\}$.

The slopes of the coordinate axes are contained in $F \cup \{\infty\}$. Hence indicator sets are always graphs of bijective mappings. This is no longer true for $L_1$-indicator sets. In general, they are not even representable as graphs of mappings, cf. section 4.1. But each $L_1$-indicator set has the same cardinality as $L$ and hence can be parametrized by the elements of $L$. Let now $\mathcal{J}$ be a $L_1$-indicator set in $L^2$. Then we have $\mathcal{J} = \{(\varphi(m), \psi(m)) \mid m \in L\}$ for suitable mappings $\varphi, \psi : L \to L$. Moreover, we can assume that $\Phi : L \to L \times L : m \mapsto (\varphi(m), \psi(m))$ is injective. Note that $\varphi$ and $\psi$ are not uniquely determined since for every bijection $\varrho : L \to L$ the mappings $\varphi \circ \varrho$ and $\psi \circ \varrho$ also parametrize $\mathcal{J}$. Assume now that two mappings $\varphi, \psi : L \to L$ are given. Then the question arises under which conditions $\varphi$ and $\psi$ define an $L_1$-indicator set in $L^2$. The answer is given by the following

PROPOSITION 2.5. *Let $F$ be a commutative field and let $L$ be a separable quadratic extension field of $F$. Assume that the affine plane over $L$ is coordinatized by a 2-dimensional left $L$-vector space. Furthermore, let $\varphi, \psi : L \to L$ be two mappings such that $\Phi : L \to L \times L : m \mapsto (\varphi(m), \psi(m))$ is injective. Then $\mathcal{J} = \{(\varphi(m), \psi(m)) \mid m \in L\}$ is an $L_1$-indicator set if and only if the following two conditions are satisfied:*

(C1) *For all $m, n \in L$ with $m \neq n$ we have $N(\varphi(m) - \varphi(n)) \neq N(\psi(m) - \psi(n))$.*

(C2) *For all $z \in L$ with $N(z) = 1$ the mapping $\sigma_z : L \to L : m \mapsto \varphi(m) - \psi(m)z$ is surjective.*

*Proof.* Let $m, n \in L$ with $m \neq n$. Since $\Phi$ is injective, the points $(\varphi(m), \psi(m))$ and $(\varphi(n), \psi(n))$ of the affine plane $L^2$ are distinct. The slope of the line connecting these points is given by $(\varphi(n) - \varphi(m))^{-1}(\psi(n) - \psi(m)) \in L \cup \{\infty\}$. The norm of this slope equals 1 if and only if we have $N(\varphi(m) - \varphi(n)) = N(\psi(m) - \psi(n))$.

Hence (C1) is equivalent to the condition we get from (I1) if we replace $F \cup \{\infty\}$ by $L_1$.

It remains to show that (C2) is equivalent to the condition that every line whose slope has norm 1 contains an element of $\mathcal{J}$. Let $z \in L$ with $N(z) = 1$. Assume that for each $b \in L$ there is at least one $m \in L$ such that the point $(\varphi(m), \psi(m))$ is on the line with slope $z$ and $y$-intercept $b$. Then the equation $\psi(m) = \varphi(m)z + b$ has at least one solution $m \in L$. Hence, for each $b \in L$ the equation $\varphi(m) - \psi(m)\overline{z} = -b\overline{z}$ has at least one solution $m \in L$. This last condition is satisfied if and only if the mapping $\sigma_{\overline{z}}$ is surjective. Since the mapping $\,^-$ leaves the unit circle invariant, we get (C2). $\square$

Obviously, the last conclusion can be reversed.

For an element $\mu \in L \setminus F$ with $N(\mu) = 1$ we define the *Cayley transformation* $\mathcal{C}_\mu : L^2 \to L^2$ by

$$\mathcal{C}_\mu(z, w) = (\overline{\mu} - \mu)^{-1} \cdot (z\overline{\mu} - w, -z\mu + w).$$

The Cayley transformation maps lines of slope $m \in L \cup \{\infty\}$ to lines of slope $(\overline{\mu} - m)^{-1}(-\mu + m)$. For calculations with $\infty$ we assume the usual conventions. Let $\hat{\mathcal{C}}_\mu$ denote the mapping induced by $\mathcal{C}_\mu$ on the projective line $L \cup \{\infty\}$. One easily calculates that $\hat{\mathcal{C}}_\mu$ maps the slopes $0, 1$ and $\infty$ to $-\mu^2, \mu$ and $-1$. Since in the inversive plane three different points are on a unique circle, it follows that $\hat{\mathcal{C}}_\mu(F \cup \{\infty\}) = L_1$. Hence $\mathcal{C}_\mu$ induces a bijection between the indicator sets and the $L_1$-indicator sets of the affine plane $L^2$. The inverse of $\mathcal{C}_\mu$ is given by $\mathcal{C}_\mu^{-1}(z, w) = (z + w, z\mu + w\overline{\mu})$.

Assume now that $\mathcal{J} = \{(\varphi(m), \psi(m)) \mid m \in L\}$ is an $L_1$-indicator set. Then $\mathcal{C}_\mu^{-1}(\mathcal{J}) = \{(\varphi(m) + \psi(m), \varphi(m)\mu + \psi(m)\overline{\mu}) \mid m \in L\}$ is an indicator set in the affine plane $L^2$. We can get a coordinate description of the spread associated with $\mathcal{J}$ by using the results of section 2.2. The elements $1$ and $\mu$ form a basis of the right $F$-vector space $L$ and $L \times L$ is a 4-dimensional right $F$-vector space. Just like in section 2.2 we put

$$S = \{0\} \times L, \quad E_1 = F \times L, \quad E_2 = \mu F \times L \quad \text{and} \quad p = \langle(1 - \mu, 0)\rangle_F.$$

Furthermore, we identify the point $m \in L$ with $\langle(1, m)\rangle_F$. Then the spread associated with $\mathcal{J}$ consists of $S$ and the subspaces

$$\begin{aligned} L(\varphi(m), \psi(m)) &= \langle(1, \varphi(m) + \psi(m)), (\mu, \varphi(m)\mu + \psi(m)\overline{\mu})\rangle_F \\ &= \{(x + \mu y, \varphi(m)(x + \mu y) + \psi(m)(x + \overline{\mu}y)) \mid x, y \in F\} \\ &= \{(z, \varphi(m)z + \psi(m)\overline{z}) \mid z \in L\} \end{aligned}$$

with $m \in L$. Algebraically $L_1$-indicator sets lead to the decomposition of $F$-linear mappings of the 2-dimensional $F$-vector space $L$ into an $L$-linear and an $L$-antilinear part. Thus we get the following

**THEOREM 2.6.** *Let $F$ be a commutative field and let $L$ be a separable quadratic extension field of $F$. Furthermore, let $\varphi, \psi : L \to L$ be two functions such that*

$\Phi : L \to L \times L : m \mapsto (\varphi(m), \psi(m))$ *is injective and assume that the conditions*
*(C1) and (C2) from Proposition 2.5 are satisfied. We put* $S = \{0\} \times L$ *and*

$$L(m, n) = \{(z, mz + n\overline{z}) \mid z \in L\} \quad \text{for } m, n \in L.$$

*Then* $\mathcal{B} = \{S\} \cup \{L(\varphi(m), \psi(m)) \mid m \in L\}$ *is a spread of the 4-dimensional right*
*F-vector space* $L^2$. *Conversely, every spread of this vector space which contains*
*the subspace $S$ admits such a description.*

*Proof.* The proof follows from Proposition 2.5 and the subsequent considerations.
Nevertheless, we give another proof which is independent of the theory developed
thus far.

Let $\mathcal{B}$ be a spread of the $F$-vector space $L^2$ with $S \in \mathcal{B}$. Every element of
$\mathcal{B} \setminus \{S\}$ is the graph of an $F$-linear mapping $\lambda : L \to L$. By Dedekind's lemma
[62: 4.14] the field automorphisms $\mathrm{id}_L$ and $\overline{\phantom{x}}$ are linearly independent over $L$.
Thus, every $F$-linear mapping $\lambda : L \to L$ can be written as

$$\lambda_{a,b} : L \to L : z \mapsto az + b\overline{z},$$

where $a, b \in L$ are uniquely determined. From Hilbert's Satz 90 [62: Theorem
4.31] we conclude that the equation $-\frac{a}{b} = \frac{\overline{z}}{z}$, which is equivalent to $\lambda_{a,b}(z) = 0$,
has a solution $z \neq 0$ if and only if $N(-\frac{a}{b}) = 1$, or equivalently $N(a) = N(b)$.
Hence, $\lambda_{a,b}$ is singular if and only if $N(a) = N(b)$. Actually, one can avoid
the use of Hilbert's Satz 90 at this point by showing directly that $\det(\lambda_{a,b}) = N(a) - N(b)$.

For $m, n \in L$ we put $L(m, n) = \{(z, mz + n\overline{z}) \mid z \in L\}$. The sets $\mathcal{B} \setminus \{S\}$
and $L$ have the same cardinality. Hence, there are mappings $\varphi, \psi : L \to L$
such that $\mathcal{B} = \{S\} \cup \{L(\varphi(m), \psi(m)) \mid m \in L\}$. Moreover, we can assume that
$\Phi : L \to L \times L : m \mapsto (\varphi(m), \psi(m))$ is injective.

For $m, n \in L$ with $m \neq n$ the linear mapping $\lambda_{\varphi(m), \psi(m)} - \lambda_{\varphi(n), \psi(n)}$ is
non-singular. Hence we have $N(\varphi(m) - \varphi(n)) \neq N(\psi(m) - \psi(n))$. It follows
that (C1) and (L1) are equivalent.

Each point $(z, w) \in L^2$ with $z \neq 0$ is contained in a subspace $L(\varphi(m), \psi(m))$
for some $m \in L$. Hence, the equation $w = \varphi(m)z + \psi(m)\overline{z}$ has at least one
solution $m \in L$. This can be rephrased as follows. For all $z \in L^\times$ the mapping
$L \to L : m \mapsto \varphi(m)z + \psi(m)\overline{z}$ is surjective. Because $z \neq 0$, we can divide by $z$
without destroying the surjectivity of this mapping. Thus we get the following
equivalent condition: For all $z \in L^\times$ the mapping $L \to L : m \mapsto \varphi(m) + \psi(m)\frac{\overline{z}}{z}$ is
surjective. By Hilbert's Satz 90 we have $L_1 = \{\frac{\overline{z}}{z} \mid z \in L^\times\}$. Since $L_1$ is invariant
under multiplication by $-1$, we get that (L2) and (C2) are equivalent. $\qquad \square$

Note that $\mathcal{J} = \{(\varphi(m), \psi(m)) \mid m \in L\}$ is an $L_1$-indicator set if and only if
$\mathcal{M}(\mathcal{J}) = \{\lambda_{\varphi(m), \psi(m)} \mid m \in L\}$ is a spread set of the $F$-vector space $L$.

In Wirtinger's approach to complex analysis the real differential of a real
differentiable mapping from $\mathbb{C}$ to $\mathbb{C}$ is decomposed into its complex linear and
its complex antilinear part. The theory of $L_1$-indicator sets gives a similar
decomposition of the linear mappings in a spread set of the $F$-vector space $L$.

## 2.6 The Kinematic Mapping

There is a remarkable connection between the considerations of the last section and the *kinematic mapping* of Blaschke and Grünwald. This kinematic mapping has the following properties, cf. [14: 50 Zusatz I]. Let $\mathcal{P}$ be the 3-dimensional real projective space and choose a fixed line $S$ of $\mathcal{P}$. The kinematic mapping now associates with each line $G$ of $\mathcal{P}$ which does not intersect $S$ two points $a(G)$ and $b(G)$ in the real euclidean plane. Two lines $G_1, G_2$ of $\mathcal{P}$ intersect if and only if the distance between $a(G_1)$ and $a(G_2)$ equals the distance between $b(G_1)$ and $b(G_2)$.

We may construct the kinematic mapping as follows: We choose four mutually different planes $E_1, \ldots, E_4$ of $\mathcal{P}$ all of which contain $S$. Let $E_i'$ denote the affine plane which is obtained from $E_i$ by deleting $S, i = 1, \ldots, 4$. The affine space obtained from $\mathcal{P}$ by deleting $E_4$ is equipped with a euclidean inner product in such a way that the distance between the affine planes $E_1'$ and $E_3'$ as well as $E_3'$ and $E_2'$ equals 1. Let now $G$ be a line of $\mathcal{P}$ which does not intersect $S$. The intersection points of $G$ with $E_1'$ and $E_2'$ are projected orthogonally onto $E_3'$. Finally, the two points of $E_3'$ we get in this way are rotated around the intersection point of $G$ and $E_3'$ by $90°$.

In order to get a coordinate description of the kinematic mapping we interprete $\mathcal{P}$ as projective space over the 4-dimensional real vector space $\mathbb{C}^2$. Furthermore, we define

$$S = \{0\} \times \mathbb{C}, \quad E_1 = (1+i)\mathbb{R} \times \mathbb{C}, \quad E_2 = (1-i)\mathbb{R} \times \mathbb{C},$$

$$E_3 = \mathbb{R} \times \mathbb{C} \quad \text{and} \quad E_4 = i\mathbb{R} \times \mathbb{C}.$$

We identify the affine space obtained from $\mathcal{P}$ by deleting $E_4$ with $i\mathbb{R} \times \mathbb{C}$. The euclidean inner product is defined by the quadratic form $Q(iy, w) = y^2 + w\overline{w}$. The rotation about $90°$ around the point $c \in \mathbb{C}$ is given by $\delta_c(w) = (w - c)i + c$.

The line connecting the points $(i, a) \in E_1'$ and $(-i, b) \in E_2'$ intersects $E_3'$ in $(0, \frac{1}{2}(a + b))$. Hence, the kinematic mapping associates with this line the pair of points

$$(\delta_{\frac{1}{2}(a+b)}(a), \delta_{\frac{1}{2}(a+b)}(b)) = (a\frac{1+i}{2} + b\frac{1-i}{2}, a\frac{1-i}{2} + b\frac{1+i}{2})$$

$$= (a, b) \begin{pmatrix} \frac{1+i}{2} & \frac{1-i}{2} \\ \frac{1-i}{2} & \frac{1+i}{2} \end{pmatrix}.$$

Let now $(a, b) \in \mathbb{C}^2$. We want to find the line $G(a, b)$ whose image under the kinematic mapping is the pair $(a, b)$. To this end we compute

$$(a, b) \begin{pmatrix} \frac{1+i}{2} & \frac{1-i}{2} \\ \frac{1-i}{2} & \frac{1+i}{2} \end{pmatrix}^{-1} = (a, b) \begin{pmatrix} \frac{1-i}{2} & \frac{1+i}{2} \\ \frac{1+i}{2} & \frac{1-i}{2} \end{pmatrix}$$

$$= (a\frac{1-i}{2} + b\frac{1+i}{2}, a\frac{1+i}{2} + b\frac{1-i}{2}).$$

Now we get the line $G(a, b)$ by connecting the points

$$(i, a\frac{1-i}{2} + b\frac{1+i}{2}) \quad \text{and} \quad (-i, a\frac{1+i}{2} + b\frac{1-i}{2}).$$

Viewed projectively this means

$$G(a, b) = \{(1 + i, a\frac{1-i}{2} + b\frac{1+i}{2})x + (1 - i, a\frac{1+i}{2} + b\frac{1-i}{2})y \mid x, y \in \mathbb{R}\}$$

$$= \{(z, \frac{a}{2}\bar{z} + \frac{b}{2}z) \mid z \in \mathbb{C}\}.$$

Hence, the kinematic mapping describes the decomposition of a linear mapping of the real vector space $\mathbb{C}$ into a complex linear and a complex antilinear part. Compared to the last section, we have a factor $\frac{1}{2}$ and the roles of $a$ and $b$ are exchanged. But this is of no significance and so we may interpret $\mathbb{C}_1$-indicator sets in $\mathbb{C}^2$ as images of spreads of the 3-dimensional real projective space under the kinematic mapping.

More generally, a kinematic mapping can be defined for each 3-dimensional pappian projective space whose ground field coordinatizes a euclidean plane. According to Fisher [30] a *euclidean plane* is a separable quadratic extension field $L$ of a commutative field $F$, cf. also [65]. The distance between two points $a, b \in L$ is defined as $N(a - b)$. In case $L = F(\sqrt{-1})$ the kinematic mapping can be constructed as above. In the general case, the existence of the kinematic mapping follows from the considerations in section 2.5. In that section we have associated two points $a(G)$ and $b(G)$ in the euclidean plane $L$ with each line $G$ of $\mathcal{P}$ that does not intersect a distinguished line $S$ of $\mathcal{P}$. The fundamental property of a kinematic mapping that two lines $G_1$ and $G_2$ intersect if and only if $N(a(G_1) - a(G_2)) = N(b(G_1) - b(G_2))$ was satisfied. As a substitute for the non-definable rotations about 90° we used the Cayley transformation. The connection between $L_1$-indicator sets and kinematic mappings will be further investigated in chapter 3.

## 2.7 $A_1$-indicator Sets

Some commutative fields do not admit any commutative quadratic separable extension field, e.g. the field of complex numbers does not. Inspection of the proof of Theorem 2.6 shows that it is not really necessary to assume that $L$ is a field. One only uses the fact that $L$ admits an involutorial automorphism $^-$ whose fixed point set is $F$ such that every $F$-linear mapping from $L$ to $L$ can be written in the form $z \mapsto az + b\bar{z}$ for suitable elements $a, b \in L$. Every commutative field admits a quadratic ring extension $A(F)$ which also has this property, the so called *double numbers over* $F$. This ring is isomorphic to the direct product $F \times F$ and it is determined up to isomorphism by this property [5: II. §5.1]. The ring $A(F)$ is also isomorphic to the quotient ring

$$F[x]/((x - a)(x - b)) \quad \text{with} \quad a, b \in F, a \neq b.$$

For our purposes a convenient choice is $a = 0$ and $b = 1$. Thus we can write

$$A = A(F) = \{x + jy \mid x, y \in F\} \quad \text{where } j^2 = j.$$

The involutorial automorphism is given by

$$\overline{x + jy} = x + y - jy.$$

The *norm* of an element $z \in A$ is defined by $N(z) = z\bar{z}$ and one easily computes that $N(x + jy) = x^2 + xy$. An element $z \in A$ is invertible if and only if $N(z) \neq 0$. The non-invertible elements are precisely the scalar multiples of $j$ and $1 - j$, in particular we have $j(1 - j) = 0$.

In analogy with Theorem 2.6 we have

THEOREM 2.7.   *Let $F$ be a commutative field and let $A$ denote the ring of double numbers over $F$. Furthermore, let $\varphi, \psi : A \to A$ be two mappings such that $\Phi : A \to A \times A : m \mapsto (\varphi(m), \psi(m))$ is injective and assume that the following properties are satisfied:*

(H1) *For all $m, n \in A$ with $m \neq n$ we have $N(\varphi(m) - \varphi(n)) \neq N(\psi(m) - \psi(n))$.*

(H2) *For all $z \in A$ with $z \neq 0$ the mapping $\varrho_z : A \to A : m \mapsto \varphi(m)z + \psi(m)\bar{z}$ is surjective.*

*We put $S = \{0\} \times A$ and*

$$L(m, n) = \{(z, mz + n\bar{z}) \mid z \in A\} \quad \text{for } m, n \in A.$$

*Then $\mathcal{B} = \{S\} \cup \{L(\varphi(m), \psi(m)) \mid m \in A\}$ is a spread of the 4-dimensional $F$-vector space $A^2$. Conversely, every spread of this vector space which contains the subspace $S$ admits such a description.*

*Proof.* We show first that each linear mapping of the vector space $A$ over $F$ can be written in the form

$$\lambda_{a,b} : A \to A : z \mapsto az + b\bar{z}$$

for suitable elements $a, b \in F$. The sets $A^2$ and $\text{End}_F(A)$ are both 4-dimensional vector spaces over $F$. Obviously, $\lambda : (a, b) \mapsto \lambda_{a,b}$ is a $F$-linear mapping from the first of these vector spaces into the second one. Therefore, it is sufficient to prove that $\lambda_{a,b}$ is the zero mapping if and only if $a = b = 0$. So let us assume that $az + b\bar{z} = 0$ for all $z \in A$. For the particular choices $z = j$ and $z = 1 - j$ we get

$$aj + b(1 - j) = 0 \quad \text{and} \quad a(1 - j) + bj = 0.$$

We multiply these two equations with $j$ as well as with $1 - j$. This gives us

$$aj^2 = aj = 0 = b(1 - j)^2 = b(1 - j) = a(1 - j)^2 = a(1 - j) = bj^2 = bj.$$

These equations imply $a = b = 0$.

Next we show that

$$\det(\lambda_{a,b}) = N(a) - N(b).$$

To this end we put $a = a_1 + ja_2$ and $b = b_1 + jb_2$ and we compute the matrix of $\lambda_{a,b}$ with respect to the basis $\{1, j\}$ of $A$. From $\lambda_{a,b}(1) = a + b$ and $\lambda_{a,b}(j) = b + (a - b)j$ we conclude that this matrix looks as follows

$$\begin{pmatrix} a_1 + b_1 & b_1 \\ a_2 + b_2 & a_1 + a_2 - b_1 \end{pmatrix}.$$

The determinant of this matrix is given by $a_1^2 + a_1 a_2 - b_1^2 - b_1 b_2 = N(a) - N(b)$.

From now on the argument runs exactly as in the proof of Theorem 2.6, only at the end it is getting shorter. One can show that there holds an analogue of Hilbert's Satz 90 for the double numbers. But this is of no use for us since it leaves out the elements of norm 0. So we cannot further reduce the condition (H2) which corresponds to (L2).  □

DEFINITION 2.8. Let $F$ be a commutative field and let $A$ denote the ring of double numbers over $F$. Furthermore, let $\varphi, \psi : A \to A$ be two mappings which satisfy the conditions (H1) and (H2) from Theorem 2.7. Assume moreover that $\Phi : A \to A \times A : m \mapsto (\varphi(m), \psi(m))$ is injective. Then the set $\mathcal{J} = \{(\varphi(m), \psi(m)) \mid m \in A\} \subset A^2$ is called an $A_1$-*indicator set*.

It is possible to prove the results of this section in a manner analogous to the considerations of section 2.5. If one wants to do this one has to look at the affine plane over the ring $A(F)$ and at the Minkowski plane induced on the improper line of this affine plane, cf. e.g. Benz [5: III. §4].

# 3. Kinematic Spaces

In this chapter we give a unifying treatment of the methods developed in chapter 2. To this end we use the theory of kinematic spaces as developed by L. Bröcker [17]. One should also note the paper [94] by E. M. Schröder. However, Schröder's definition is somewhat more special than Bröcker's, so that the class of stretch translation spaces, which is of special importance for us, is not covered.

From [17] we adopt the following

DEFINITION 3.1.   A *kinematic space* is a pair $(\mathcal{P}, \mathcal{G})$, where $\mathcal{P} = (P, \mathcal{L})$ is a projective space over a skewfield $F$ and $\mathcal{G} \subseteq P$ is a group such that the following conditions are satisfied:

(KS1) For each $L \in \mathcal{L}$ with $L \cap \mathcal{G} \neq \emptyset$ we have $|L \setminus \mathcal{G}| \leq 2$.

(KS2) For all $L \in \mathcal{L}$ with $e \in L$ the set $L \cap \mathcal{G}$ is a subgroup of $\mathcal{G}$. Here $e \in \mathcal{G}$ denotes the identity.

(KS3) For all $g \in \mathcal{G}$ the left and right translations $\lambda_g, \rho_g : \mathcal{G} \to \mathcal{G}$ with $\lambda_g(x) = gx$ and $\rho_g(x) = xg$ can be extended to collineations of $\mathcal{P}$.

If $F$ is infinite, Bröcker only requires that each line which intersects $\mathcal{G}$ contains only finitely many points which are not in $\mathcal{G}$. He shows that this condition implies (KS1) [17: Satz 5]. Moreover, it is sufficient to require that the left and right translations by elements of $\mathcal{G}$ are collineations of the geometry $(\mathcal{G}, \{L \cap \mathcal{G} \mid L \in \mathcal{L}\})$. It turns out that these mappings are induced by collineations of $\mathcal{P}$ [17: Lemma 1].

For us, kinematic spaces with $\dim \mathcal{P} = 3$ are of course the most interesting ones. In this case, almost all lines of $\mathcal{P}$ are cosets of subgroups of $\mathcal{G}$. Moreover, it turns out that in most cases the group $\mathcal{G}$ naturally acts as a collineation group of a projective or affine plane in such a way that the lines through $e$ are the point stabilizers of the action. Since right multiplication by elements of $\mathcal{G}$ induce collineations of $\mathcal{P}$, the other lines of $\mathcal{P}$ are right cosets of the lines through $e$. The right cosets of the point stabilizers consist precisely of the mappings which map one given point onto another one. Hence we may use these two points to parametrize the lines of $\mathcal{P}$.

## 3.1 Stretch Translation Spaces

Let $F$ be a skewfield, $E$ the desarguesian affine plane over $F$, and $\mathcal{G}$ the dilatation group of $E$. We coordinatize $E$ using a 2-dimensional right vector space over $F$. For two points $a, b \in E$ we define

$$L(a, b) = \{\delta \in \mathcal{G} \mid \delta(a) = b\}.$$

PROPOSITION 3.2. *The group $\mathcal{G}$ can be identified with the complement of the union of two planes $E_1, E_2$ in a 3-dimensional projective space $\mathcal{P}$ over $F$ in such a way that $(\mathcal{P}, \mathcal{G})$ becomes a kinematic space. Let $S$ denote the intersection of $E_1$ and $E_2$ and let $E_i'$ be the affine plane obtained from $E_i$ by deleting $S$, $i = 1, 2$. There are collineations $\pi_i : E \to E_i', i = 1, 2$, such that the lines of $\mathcal{P}$ which do not intersect $S$ are precisely the sets $\overline{L}(a, b) = L(a, b) \cup \{\pi_1(a), \pi_2(b)\}$ where $a, b \in E$. Two lines $\overline{L}(a_1, b_1)$ and $\overline{L}(a_2, b_2)$ intersect if and only if $a_1 = a_2$ or $b_1 = b_2$ or the lines $a_1 \vee a_2$ and $b_1 \vee b_2$ are parallel in the affine plane $E$.*

*Proof.* Let $\mathcal{P} = (P, \mathcal{L})$ be the projective space over the 4-dimensional right vector space $E \times F^2$. All mappings in $\mathcal{G}$ can be written as

$$\delta_{s,t} : E \to E : x \mapsto xs + t,$$

where $s \in F^\times, t \in E$. We identify this mapping with the point $\langle(t, s, 1)\rangle$ of $\mathcal{P}$. The complement of $\mathcal{G}$ in the point set of $\mathcal{P}$ is the union of the planes

$$E_1 = \{(x, y, 0) \mid x \in E, y \in F\} \quad \text{and} \quad E_2 = \{(x, 0, z) \mid x \in E, z \in F\}.$$

We define $S, E_1'$ and $E_2'$ as given in the proposition. Moreover, we put

$$\pi_1 : E \to E_1' : x \mapsto \langle(-x, 1, 0)\rangle \quad \text{and}$$
$$\pi_2 : E \to E_2' : x \mapsto \langle(x, 0, 1)\rangle.$$

Let $a, b \in E$, then we have $L(a, b) = \{\delta_{s,t} \in \mathcal{G} \mid t = b - as\}$. The corresponding set in $P$ is $\{\langle(b - as, s, 1)\rangle \mid s \in F^\times\}$. The union of this set with $\langle(b, 0, 1)\rangle = \pi_2(b)$ and $\langle(-a, 1, 0)\rangle = \pi_1(a)$ is a line of $\mathcal{P}$ which does not intersect $S$. Conversely, every line of $\mathcal{P}$ which does not intersect $S$ can be written this way.

Let now $\overline{L}(a_1, b_1)$ and $\overline{L}(a_2, b_2)$ be two intersecting lines of $\mathcal{P}$.

If the intersection point of these lines is in $E_1'$ or $E_2'$, respectively, then we have $a_1 = a_2$ or $b_1 = b_2$.

In case $\overline{L}(a_1, b_1) \cap \overline{L}(a_2, b_2) \in \mathcal{G}$ there is an element $\delta \in \mathcal{G}$ such that $\delta(a_1) = b_1$ and $\delta(a_2) = b_2$. Such a dilatation $\delta$ exists if and only if the lines $a_1 \vee a_2$ and $b_1 \vee b_2$ are parallel in the affine plane $E$. $\square$

Assume now that $\mathcal{B}$ is a spread of $\mathcal{P}$ such that $S \in \mathcal{B}$. Since each point of $E_1'$ and $E_2'$ is contained in exactly one element of $\mathcal{B} \setminus \{S\}$, there exists a bijective mapping $f : E \to E$ such that $\mathcal{B} \setminus \{S\} = \{\overline{L}(a, f(a)) \mid a \in E\}$. Different elements of $\mathcal{B}$ do not intersect. Hence, for any $a, b \in E$ with $a \neq b$, the lines $a \vee b$ and $f(a) \vee f(b)$ are not parallel in the affine plane $E$. Since through each point of $\mathcal{P}$ which is not contained in $E_1 \cup E_2$ there passes at least one element of $\mathcal{B} \setminus \{S\}$, there exists for each dilatation $\delta \in \mathcal{G}$ at least one $a \in E$ with $\delta(a) = f(a)$. Thus, $f$ is a transversal mapping.

## 3.2 Euclidean Motion Groups

Let $F$ be a commutative field and let $L$ be a *euclidean plane over $F$*. This means that $L$ is a separable quadratic extension field of $F$, cf. section 2.5. The distance betweeen two points $a, b \in L$ is defined as $N(a-b)$. The *proper euclidean motion group* $\mathcal{G}$ consists of all mappings $\sigma_{c,t} : L \to L : z \mapsto zc + t$ where $c, t \in L$ and $N(c) = 1$. For $a, b \in L$ we put

$$L(a, b) = \{\sigma \in \mathcal{G} \mid \sigma(a) = b\}.$$

PROPOSITION 3.3. *The group $\mathcal{G}$ can be identified with the complement of a line $S$ in a 3-dimensional projective space $\mathcal{P}$ over $F$ in such a way that $(\mathcal{P}, \mathcal{G})$ becomes a kinematic space. The lines of $\mathcal{P}$ which do not intersect $S$ are precisely the sets $L(a, b)$ where $a, b \in L$. Two lines $L(a_1, b_1)$ and $L(a_2, b_2)$ intersect if and only if $N(a_1 - a_2) = N(b_1 - b_2)$.*

*Proof.* Let $\mathcal{P} = (P, \mathcal{L})$ be the projective space associated with the 4-dimensional $F$-vector space $L^2$ and put $S = \{0\} \times L \in \mathcal{L}$. We want to construct a bijection $\Psi : \mathcal{G} \to P \setminus S$. Let $c, t \in L$ with $N(c) = 1$. By Hilbert's Satz 90, there exists $z \in L$ such that $c = \frac{\bar{z}}{z}$. We define $\Psi(\sigma_{c,t}) = \langle (z, tz) \rangle$. Note that if $\frac{\bar{z}}{z} = \frac{\bar{w}}{w}$ then $z$ and $w$ are linearly dependent over $F$. Hence $\Psi$ is well defined.

Since $t \in L$ and $z \in L \setminus \{0\}$ can be chosen arbitrary, the mapping $\Psi$ is surjective.

Let $c_i, t_i \in L, N(c_i) = 1, i = 1, 2$, and assume that $\Psi(\sigma_{c_1, t_1}) = \Psi(\sigma_{c_2, t_2})$. Let $z_i \in L$ such that $c_i = \frac{\bar{z}_i}{z_i}, i = 1, 2$. Then $(z_1, t_1 z_1)$ and $(z_2, t_2 z_2)$ are linearly dependent over $F$. Hence, $z_1$ and $z_2$ are linearly dependent as well and we get $c_1 = c_2$. Since $z_i \neq 0$, we also get $t_1 = t_2$. Thus, $\Psi$ is injective.

Let $a, b \in L$. In order to find $L(a, b)$ we have to solve the equation $\sigma_{c,t}(a) = ac + t = b$. This yields $t = b - ac$ and hence we get $\Psi(L(a, b)) = \{\langle (z, bz - a\bar{z}) \rangle \mid z \in L \setminus \{0\}\}$. This shows that $\Psi(L(a, b))$ is a line of $\mathcal{P}$ that does not intersect $S$, and since $a$ and $b$ are arbitrary elements of $L$, each of these lines admits such a description.

Two lines $L(a_1, b_1)$ and $L(a_2, b_2)$ intersect if and only if there exists a motion $\sigma \in \mathcal{G}$ such that $\sigma(a_1) = b_1$ and $\sigma(a_2) = b_2$. Such a motion exists if and only if $N(a_1 - a_2) = N(b_1 - b_2)$. $\square$

In the sequel we omit the mapping $\Psi$ and identify $\mathcal{G}$ and $P \setminus S$.

Let $\mathcal{B}$ be a spread of $\mathcal{P}$ with $S \in \mathcal{B}$. Since $|\mathcal{B} \setminus \{S\}| = |L|$, there exist mappings $\varphi, \psi : L \to L$ such that $\mathcal{B} \setminus \{S\} = \{L(\varphi(m), \psi(m)) \mid m \in L\}$ and $\Phi : L \to L \times L : m \mapsto (\varphi(m), \psi(m))$ is injective. Since different elements of $\mathcal{B} \setminus \{S\}$ do not intersect, we have $N(\varphi(m) - \varphi(n)) \neq N(\psi(m) - \psi(n))$ for $m \neq n$. So $\varphi$ and $\psi$ satisfy condition (C1) of Proposition 2.5. Through each $\sigma \in \mathcal{G}$ there passes at least one element of $\mathcal{B} \setminus \{S\}$. Hence, the equation $\sigma(\varphi(m)) = \psi(m)$ has a solution $m \in L$. The algebraic description of $\mathcal{G}$ shows that this last condition is equivalent to (C2). So $\{(\varphi(m), \psi(m)) \mid m \in L\}$ is an $L_1$-indicator set in the affine plane $L^2$.

## 3.3 Minkowskian Motion Groups

This section is based on the paper [93] by E. M. Schröder.

Let $F$ be a commutative field and let $A = A(F)$ be a *Minkowskian plane over $F$*. This means that $A$ is isomorphic to the ring of double numbers over $F$, cf. section 2.7. The distance between two points $a, b \in A$ is given by $N(a - b)$. Moreover, $A$ is an affine plane over $F$. The *proper Minkowskian motion group $\mathcal{G}$* of $A$ consists of all mappings $\sigma : A \to A : z \mapsto zc + t$ with $c, t \in A$ and $N(c) = 1$. For $a, b \in A$ we define

$$L(a, b) = \{\sigma \in \mathcal{G} \mid \sigma(a) = b\}.$$

For $A$ we use the same description as in section 2.7. Then we have the following

PROPOSITION 3.4. *The group $\mathcal{G}$ can be identified with the complement of the union of two planes $E_1, E_2$ in a 3-dimensional projective space $\mathcal{P}$ over $F$ in such a way that $(\mathcal{P}, \mathcal{G})$ becomes a kinematic space. Let $S$ denote the intersection of $E_1$ and $E_2$ and let $E_i'$ be the affine plane obtained from $E_i$ by deleting $S, i = 1, 2$. There are collineations $\pi_i : A \to E_i', i = 1, 2$, such that the lines of $\mathcal{P}$ which do not intersect $S$ are precisely the sets $\overline{L}(a, b) = L(a, b) \cup \{\pi_1(ja + (1-j)b), \pi_2((1-j)a + jb)\}$ with $a, b \in A$. Two lines $\overline{L}(a_1, b_1)$ and $\overline{L}(a_2, b_2)$ intersect if and only if $N(a_1 - a_2) = N(b_1 - b_2)$.*

*Proof.* Let $\mathcal{P}$ be the projective space associated with the right vector space $F^4$. The elements $j$ and $1 - j$ form a basis of the right $F$-vector space $A$. The mappings in $\mathcal{G}$ can be written as

$$jx + (1 - j)y \mapsto j\alpha^{-1}(x + \beta) + (1 - j)(\alpha y + \gamma),$$

where $\alpha, \beta, \gamma \in F, \alpha \neq 0$ [93: Satz 1.1]. We identify this mapping with the point $\langle (\alpha, \beta, \gamma, 1) \rangle$ of $\mathcal{P}$. The pair $(\mathcal{P}, \mathcal{G})$ then is a kinematic space [93: Satz 2.2]. The complement of $\mathcal{G}$ in the point set of $\mathcal{P}$ is the union of the planes

$$E_1 = \{(0, y, u, v) \mid y, u, v \in F\} \quad \text{and} \quad E_2 = \{(x, y, u, 0) \mid x, y, u \in F\}.$$

We define $S, E_1'$ and $E_2'$ as in the proposition. Moreover, we put

$$\pi_1 : A \to E_1' : jx + (1 - j)y \mapsto \langle (0, -x, y, 1) \rangle \quad \text{and}$$
$$\pi_2 : A \to E_2' : jx + (1 - j)y \mapsto \langle (1, x, -y, 0) \rangle.$$

Let $a = ja_1 + (1-j)a_2, b = jb_1 + (1-j)b_2 \in A$, where $a_i, b_i \in F$. The line joining $\pi_1(ja + (1-j)b) = \pi_1(ja_1 + (1-j)b_2)$ and $\pi_2((1-j)a + jb) = \pi_2(jb_1 + (1-j)a_2)$ in $\mathcal{P}$ consists of the points

$$\langle (0, -a_1, b_2, 1)\lambda + (1, b_1, -a_2, 0)\mu \rangle \quad \text{with} \quad \lambda, \mu \in F.$$

Such a point is contained in $\mathcal{G}$ if and only if $\lambda \neq 0 \neq \mu$. If this holds, then the mapping corresponding to this point is given by

$$jx + (1-j)y \mapsto j\frac{\lambda}{\mu}(x - a_1 + \frac{\mu}{\lambda}b_1) + (1-j)(\frac{\mu}{\lambda}y + b_2 - \frac{\mu}{\lambda}a_2).$$

These are precisely the elements of $\mathcal{G}$ that map $a$ onto $b$.

Let now $\overline{L}(a_1, b_1)$ and $\overline{L}(a_2, b_2)$ be two intersecting lines of $\mathcal{P}$.

If the intersection point of these lines is in $E_1'$ or $E_2'$, then we have $ja_1 + (1-j)b_1 = ja_2 + (1-j)b_2$ or $(1-j)a_1 + jb_1 = (1-j)a_2 + jb_2$, respectively. If we multiply the first of these equations by $j$ and $1-j$ we get $ja_1 - ja_2 = 0$ and $(1-j)b_1 - (1-j)b_2 = 0$. Hence, $a_1 - a_2$ and $b_1 - b_2$ are zero divisors, which implies that $N(a_1 - a_2) = N(b_1 - b_2) = 0$. A similar reasoning applies to the second equation.

In case $\overline{L}(a_1, b_1) \cap \overline{L}(a_2, b_2) \in \mathcal{G}$ there is an element $\sigma \in \mathcal{G}$ such that $\sigma(a_1) = b_1$ and $\sigma(a_2) = b_2$. Thus we have $N(a_1 - a_2) = N(b_1 - b_2)$.

Let now $a_1, a_2, b_1, b_2 \in A$ such that $N(a_1 - a_2) = N(b_1 - b_2)$. We must show that the lines $\overline{L}(a_1, b_1)$ and $\overline{L}(a_2, b_2)$ intersect.

In case $N(a_1 - a_2) \neq 0$ there exists $\sigma \in \mathcal{G}$ such that $\sigma(a_1) = b_1$ and $\sigma(a_2) = b_2$. Hence, the lines $\overline{L}(a_1, b_1)$ and $\overline{L}(a_2, b_2)$ intersect in $\mathcal{G}$.

Let now $N(a_1 - a_2) = 0$, then we have $a_1 - a_2, b_1 - b_2 \in jF \cup (1-j)F$, cf. section 2.7. If both of them are contained in $jF$ or $(1-j)F$, then there is an element $\sigma \in \mathcal{G}$ with $\sigma(a_1) = b_1$ and $\sigma(a_2) = b_2$. Consequently, $\overline{L}(a_1, b_1)$ and $\overline{L}(a_2, b_2)$ intersect in $\mathcal{G}$. In case $a_1 - a_2 \in jF$ and $b_1 - b_2 \in (1-j)F$ we have $(1-j)(a_1 - a_2) = 0 = -j(b_1 - b_2)$. Hence, the lines $\overline{L}(a_1, b_1)$ and $\overline{L}(a_2, b_2)$ have an intersection point in $E_2'$. Similarly, if $a_1 - a_2 \in (1-j)F$ and $b_1 - b_2 \in jF$, then $\overline{L}(a_1, b_1)$ and $\overline{L}(a_2, b_2)$ have an intersection point in $E_1'$. $\square$

Let us now consider a spread $\mathcal{B}$ of $\mathcal{P}$ with $S \in \mathcal{B}$. Since $|\mathcal{B} \setminus \{S\}| = |A|$, there are mappings $\varphi, \psi : A \to A$ such that $\mathcal{B} \setminus \{S\} = \{\overline{L}(\varphi(m), \psi(m)) \mid m \in A\}$ and $\Phi : A \to A \times A : m \mapsto (\varphi(m), \psi(m))$ is injective. Since different elements of $\mathcal{B} \setminus \{S\}$ do not intersect, we have $N(\varphi(m) - \varphi(n)) \neq N(\psi(m) - \psi(n))$ for $m \neq n$. Hence, the mappings $\varphi$ and $\psi$ satisfy the condition (H1) from Theorem 2.7.

Through each $\sigma \in \mathcal{G}$ there passes at least one element of $\mathcal{B} \setminus \{S\}$. So, the equation $\sigma(\varphi(m)) = \psi(m)$ has at least one solution $m \in A$. This implies that the mapping $\chi_c : A \to A : m \mapsto \varphi(m)c - \psi(m)$ is surjective for all $c \in A$ with $N(c) = 1$. Let $z \in A$ with $N(z) \neq 0$. Then we have $N(-\frac{\overline{z}}{z}) = 1$. We put $c = -\frac{\overline{z}}{z}$ and multiply $\chi_c$ by $-\overline{z}$. This shows that the mapping $\varrho_z : A \to A : m \mapsto \varphi(m)z + \psi(m)\overline{z}$ is surjective as well. Since through each point of $E_1'$ and $E_2'$ there also passes at least one element of $\mathcal{B} \setminus \{S\}$, the mappings $\varrho_z, z \in A$

with $N(z) = 0$, are surjective as well. So, the mappings $\varphi$ and $\psi$ also satisfy the condition (H2) from Theorem 2.7 and hence $\{(\varphi(m), \psi(m)) \mid m \in A\}$ is an $A_1$-indicator set.

## 3.4 Elliptic Motion Groups

Let $F$ be a commutative field and let $\mathbf{H}$ be a quaternion skewfield over $F$ [84: 6.3]. We put $\mathcal{G} = \mathbf{H}^{\times}/F^{\times}$ and we identify $\mathcal{G}$ with the point set of the 3-dimensional projective space associated with the $F$-vector space $\mathbf{H}$. Since all 2-dimensional $F$-subspaces of $\mathbf{H}$ which contain 1 are subalgebras of $\mathbf{H}$, the pair $(\mathcal{P}, \mathcal{G})$ is a kinematic space. For this reason, $\mathbf{H}$ is sometimes called a kinematic algebra, cf. [66].

The algebra $\mathbf{H}$ admits an involutorial antiautomorphism $^-$ and a quadratic norm form $N : \mathbf{H} \to F : z \mapsto z\bar{z}$ with associated bilinear form $f : \mathbf{H} \times \mathbf{H} \to F : (z, w) \mapsto z\bar{w} + w\bar{z}$. We put $\operatorname{Pu} \mathbf{H} = \{z \in \mathbf{H} \mid z + \bar{z} = 0\}$. Then $\operatorname{Pu} \mathbf{H}$ is a 3-dimensional subspace of $\mathbf{H}$ and the group $\mathcal{G}$ operates on $\operatorname{Pu} \mathbf{H}$ by conjugation. $\mathcal{G}$ is called the *elliptic motion group* of the projective plane associated with $\operatorname{Pu} \mathbf{H}$.

If one tries to use the elliptic motion group for the description of spreads of $\mathcal{P}$, one encounters several difficulties. The first one is that $\mathcal{G}$ does not necessarily operate transitively on the projective plane over $\operatorname{Pu} \mathbf{H}$. Secondly the stabilizer of the point $\langle z \rangle < \operatorname{Pu} \mathbf{H}$ is the union of the lines $\langle 1, z \rangle$ and $\operatorname{Pu} \mathbf{H} \cap (\operatorname{Pu} \mathbf{H}) \cdot z$ [73: remark on p. 68; 17: Satz 8]. These two lines coincide if and only if the characteristic of $F$ equals 2.

In order to interpret the lines of $\mathcal{P}$ as cosets of point stabilizers one therefore should consider an action of $\mathcal{G}$ on a different set. This works in case of the real quaternion skewfield $\mathbb{H}$. It turns out that the action of $\mathcal{G}$ on the set $\operatorname{SPu} \mathbb{H} = \{z \in \operatorname{Pu} \mathbb{H} \mid N(z) = 1\}$ leads to appropriate results.

The pair $(\operatorname{Pu} \mathbb{H}, f)$ is nothing else than the 3-dimensional real euclidean space, $\operatorname{SPu} \mathbb{H}$ is the unit sphere of this space, and $\mathcal{G}$ operates as the rotation group of this space, cf. Blaschke [14: 75].

We define the *spherical distance* between two points $z, w \in \operatorname{SPu} \mathbb{H}$ by $d(z, w) = \arccos(\frac{1}{2} f(z, w))$, where we assume this number to be contained in the interval $[0, \pi]$. For $z, w \in \operatorname{SPu} \mathbb{H}$ we put

$$L(z, w) = \{\sigma \in \mathcal{G} \mid \sigma(z) = w\}.$$

The sets $L(z, w), z, w \in \operatorname{SPu} \mathbb{H}$, are the lines of $\mathcal{P}$. The lines $L(z_1, w_1)$ and $L(z_2, w_2)$ coincide if and only if $(z_1, w_1) = \pm(z_2, w_2)$. Moreover, two lines intersect if and only if $d(z_1, z_2) = d(w_1, w_2)$, cf. [14: 76]. The mapping $(z, w) \mapsto L(z, w)$ is called *Hjelmslev's line mapping* by Blaschke. Since $\operatorname{SPu} \mathbb{H} \times \operatorname{SPu} \mathbb{H}$ is simply connected, we can view this mapping as the universal covering of the line space of $\mathcal{P}$.

A mapping $\varphi : \operatorname{SPu} \mathbb{H} \to \operatorname{SPu} \mathbb{H}$ is called a *contraction* if we have $d(\varphi(z), \varphi(w)) < d(z, w)$ for all $z, w \in \operatorname{SPu} \mathbb{H}$ with $z \neq w$.

PROPOSITION 3.5. *Let $\mathbb{H}$ be the skewfield of real quaternions, and let $\mathcal{B}$ be a topological spread of the projective space $\mathcal{P}$ associated with the 4-dimensional real vector space $\mathbb{H}$. Then there exists a contraction $\varphi : \mathbb{S}\mathrm{Pu}\,\mathbb{H} \to \mathbb{S}\mathrm{Pu}\,\mathbb{H}$ such that $\mathcal{B}$ consists either of all lines $L(m, \varphi(m)), m \in \mathbb{S}\mathrm{Pu}\,\mathbb{H}$, or of all lines $L(\varphi(m), m), m \in \mathbb{S}\mathrm{Pu}\,\mathbb{H}$. Conversely, if $\varphi : \mathbb{S}\mathrm{Pu}\,\mathbb{H} \to \mathbb{S}\mathrm{Pu}\,\mathbb{H}$ is a contraction, then $\mathcal{B} = \{L(m, \varphi(m)) \,|\, m \in \mathbb{S}\mathrm{Pu}\,\mathbb{H}\}$ and $\mathcal{B}' = \{L(\varphi(m), m) \,|\, m \in \mathbb{S}\mathrm{Pu}\,\mathbb{H}\}$ are topological spreads of $\mathcal{P}$. These two spreads are equivalent.*

*Proof.* Let $\mathcal{B}$ be a topological spread of the projective space associated with $\mathbb{H}$. According to Proposition 1.27, $\mathcal{B}$ is homeomorphic to $\mathbb{S}_2$. Since $\mathbb{S}_2$ is simply connected, the preimage of $\mathcal{B}$ under Hjelmslev's line mapping is the disjoint union of two sets $\mathcal{B}_1 \subset \mathbb{S}\mathrm{Pu}\,\mathbb{H} \times \mathbb{S}\mathrm{Pu}\,\mathbb{H}$ and $\mathcal{B}_2 = -\mathcal{B}_1$, which are both homeomorphic to $\mathbb{S}_2$. Let $\pi_1, \pi_2 : \mathbb{S}\mathrm{Pu}\,\mathbb{H} \times \mathbb{S}\mathrm{Pu}\,\mathbb{H} \to \mathbb{S}\mathrm{Pu}\,\mathbb{H}$ denote the canonical projections.

For any two distinct elements $(m_1, n_1), (m_2, n_2) \in \mathcal{B}_1$ we have $d(m_1, m_2) \neq d(n_1, n_2)$. Since $\mathcal{B}_1$ is simply connected and $d$ is continuous, this implies that for all $(m_1, n_1), (m_2, n_2) \in \mathcal{B}_1$ with $(m_1, n_1) \neq (m_2, n_2)$ either

$$(I) \qquad\qquad d(m_1, m_2) < d(n_1, n_2)$$

or

$$(II) \qquad\qquad d(m_1, m_2) > d(n_1, n_2).$$

In case (I) the restriction of $\pi_2$ to $\mathcal{B}_1$ is injective and by Brouwer's theorem it is even a homeomorphism. Moreover, there exists a contraction $\varphi : \mathbb{S}\mathrm{Pu}\,\mathbb{H} \to \mathbb{S}\mathrm{Pu}\,\mathbb{H}$ such that $\mathcal{B}_1 = \{(\varphi(m), m) \,|\, m \in \mathbb{S}\mathrm{Pu}\,\mathbb{H}\}$. An analogous reasoning applies to case (II).

Assume now that $\varphi : \mathbb{S}\mathrm{Pu}\,\mathbb{H} \to \mathbb{S}\mathrm{Pu}\,\mathbb{H}$ is a contraction. Then the elements of $\mathcal{B} = \{L(m, \varphi(m)) \,|\, m \in \mathbb{S}\mathrm{Pu}\,\mathbb{H}\}$ are mutually complementary and $\mathcal{B}$ is homeomorphic to $\mathbb{S}_2$. By Proposition 1.27 $\mathcal{B}$, is a topological spread. By the same argument, $\mathcal{B}' = \{L(\varphi(m), m) \,|\, m \in \mathbb{S}\mathrm{Pu}\,\mathbb{H}\}$ is also a topological spread.

It remains to show that $\mathcal{B}$ and $\mathcal{B}'$ are equivalent. By [17: Satz 1], inversion in $\mathcal{G}$ induces a collineation of $\mathcal{P}$. Let $m, n \in \mathbb{S}\mathrm{Pu}\,\mathbb{H}$. By definition we have $L(m, n) = \{z\mathbb{R}^\times \in \mathcal{G} \,|\, z^{-1}mz = n\}$. This shows that inversion in $\mathcal{G}$ interchanges the lines $L(m, n)$ and $L(n, m)$. It follows immediately that $\mathcal{B}$ and $\mathcal{B}'$ are equivalent. $\qquad\square$

The fact that topological spreads of a 4-dimensional real vector space can be described using contractions of the 2-sphere was already proved by Gluck und Warner [34: Theorem A]. However, the use of the quaternions simplifies and shortens their proof considerably.

## 3.5 Other Kinematic Spaces

Apart from the 3-dimensional kinematic spaces considered up to now there are the group spaces of hyperbolic motion groups and the so called degenerate kinematic spaces.

Each hyperbolic motion group $\mathcal{G}$ can be identified with the complement of a ruled quadric $\mathcal{H}$ in a 3-dimensional projective space $\mathcal{P}$ in such a way that $(\mathcal{P},\mathcal{G})$ becomes a kinematic space [94: §7]. If one wants to use $\mathcal{G}$ for the description of a spread $\mathcal{B}$ of $\mathcal{P}$ one has to take into account the mutual relationship between $\mathcal{B}$ and $\mathcal{H}$. It is conceivable that one can develop methods similar to the ones discussed up to now for the description of spreads which contain one of the reguli contained in $\mathcal{H}$.

The degenerate kinematic spaces admit a description as motion groups. However, it is not possible to interpret the lines as point stabilizers.

# 4. Examples and Supplements

## 4.1 André Planes

Let $F$ be a commutative field and let $L$ be a separable quadratic extension field of $F$. Furthermore, let $N : L \to F$ denote the norm.

Let $M \subset F$. We define two functions $\varphi, \psi : L \to L$ by

$$\varphi(m) = \begin{cases} m & \text{if } N(m) \in M \\ 0 & \text{if } N(m) \notin M, \end{cases} \qquad \psi(m) = \begin{cases} 0 & \text{if } N(m) \in M \\ m & \text{if } N(m) \notin M. \end{cases}$$

Obviously, $\Phi : F \to F \times F : m \mapsto (\varphi(m), \psi(m))$ is injective. Let $m, n \in L$ with $m \neq n$. Then we have

$$N(\varphi(m) - \varphi(n)) = \begin{cases} N(m - n) & \text{if } N(m), N(n) \in M \\ N(m) & \text{if } N(m) \in M, N(n) \notin M \\ N(n) & \text{if } N(m) \notin M, N(n) \in M \\ 0 & \text{if } N(m), N(n) \notin M \end{cases}$$

and

$$N(\psi(m) - \psi(n)) = \begin{cases} 0 & \text{if } N(m), N(n) \in M \\ N(n) & \text{if } N(m) \in M, N(n) \notin M \\ N(m) & \text{if } N(m) \notin M, N(n) \in M \\ N(m - n) & \text{if } N(m), N(n) \notin M. \end{cases}$$

In any case we have $N(\varphi(m) - \varphi(n)) \neq N(\psi(m) - \psi(n))$. Hence, $\varphi$ and $\psi$ satisfy condition (C1) from Proposition 2.5.

Let $z \in L$ with $N(z) = 1$. We want to show that $\sigma_z : L \to L : m \mapsto \varphi(m) - \psi(m)z$ is surjective. We have

$$\sigma_z(m) = \begin{cases} m & \text{if } N(m) \in M \\ -mz & \text{if } N(m) \notin M, \end{cases}$$

and since $N(z) = 1$, it follows that $\sigma_z$ is surjective. Hence, $\varphi$ and $\psi$ also satisfy condition (C2) from Proposition 2.5 and $\mathcal{J} = \{(\varphi(m), \psi(m)) \mid m \in L\} \subset L^2$ is an $L_1$-indicator set. The translation plane associated with $\mathcal{J}$ is called an *André plane* [1: §8.2].

## 4.2 Remarks on the Collineation Group

The question whether two different $L_1$- or $A_1$-indicator sets define isomorphic translation planes is not easy to decide in general. This is also true for the related question what the collineation group of the translation plane associated with an $L_1$- or $A_1$-indicator set looks like. However, it is possible to get a satisfactory description of the collineations which fix the distinguished subspace $S$.

In the sequel let $F$ be a commutative field and let $E$ denote either a separable, quadratic extension field of $F$ or the ring of double numbers over $F$. Every $F$-linear mapping $\lambda : E \to E$ can be written as

$$\lambda_{a,b} : E \to E : z \mapsto az + b\bar{z}$$

where $a, b \in E$ are uniquely determined.

LEMMA 4.1. *Let* $a_i, b_i \in E, i = 1, 2$, *then we have*

$$\lambda_{a_1,b_1} \circ \lambda_{a_2,b_2} = \lambda_{a_1 a_2 + b_1 \bar{b}_2, a_1 b_2 + b_1 \bar{a}_2}.$$

*Let* $a, b \in E$. *Then* $\lambda_{a,b}$ *is invertible if and only if* $N(a) - N(b) = \det(\lambda_{a,b}) \neq 0$. *If this holds then we have*

$$\lambda_{a,b}^{-1} = \frac{1}{N(a) - N(b)} \lambda_{\bar{a}, -b}.$$

*Proof.* We have

$$\lambda_{a_1,b_1}(\lambda_{a_2,b_2}(z)) = a_1(a_2 z + b_2 \bar{z}) + b_1(\bar{a}_2 \bar{z} + \bar{b}_2 z) = \lambda_{a_1 a_2 + b_1 \bar{b}_2, a_1 b_2 + b_1 \bar{a}_2}(z),$$

which proves the first assertion. As an application we get

$$\lambda_{a,b} \circ \lambda_{\bar{a},-b} = \lambda_{N(a)-N(b),0} = (N(a) - N(b)) \cdot \mathrm{id}_E,$$

and the second assertion follows. The proof that $\det(\lambda_{a,b}) = N(a) - N(b)$ can be obtained as in Theorem 2.7. $\square$

This lemma says that we may describe $\mathrm{End}_F(E)$ as a Cayley-Dickson extension of $E$, cf. [92: III.4]. It is possible to prove Theorem 2.6 and 2.7 using this lemma.

Let $\sigma : E \to E$ be a ring automorphism such that $F^\sigma = F$. Assume moreover that $\sigma$ and $\bar{\ }$ commute. In case $E$ is the ring of double numbers over $F$ each field automorphism of $F$ can be extended to an automorphism of $E$ which commutes with $\bar{\ }$ and obviously fixes $F$. In case $E$ is a separable extension field this is not true in general. However, it is true if $F$ is a finite field or the field of real numbers.

Let $\lambda : E \to E$ be semilinear with respect to $\sigma|_F$. Then there exist $a, b \in E$ such that $\lambda$ is of the form

$$\lambda_{a,b,\sigma} : E \to E : z \mapsto az^\sigma + b\bar{z}^\sigma.$$

We have $\lambda_{a,b,\sigma} = \lambda_{a,b} \circ \sigma$. Hence, $\lambda_{a,b,\sigma}$ is invertible if and only if $N(a) \neq N(b)$. If this holds true we get

$$\lambda_{a,b,\sigma}^{-1} = \sigma^{-1} \circ \left( \tfrac{1}{N(a)-N(b)} \lambda_{\bar{a},-b} \right).$$

We need the following

LEMMA 4.2. *For all $a, b \in E$ we have $\sigma \circ \lambda_{a,b} \circ \sigma^{-1} = \lambda_{a^\sigma, b^\sigma}$.*

*Proof.* Let $z \in E$. Then we have

$$(\sigma \circ \lambda_{a,b} \circ \sigma^{-1})(z) = \left( a z^{\sigma^{-1}} + b \bar{z}^{\sigma^{-1}} \right)^\sigma = (a^\sigma z + b^\sigma \bar{z}),$$

since $\sigma$ is an automorphism of $E$ which commutes with $^-$. $\qquad\square$

Let $a, b \in E$. The graph of the linear mapping $\lambda_{a,b}$ is denoted by $L(a,b)$, i.e. we put

$$L(a,b) = \{(z, az + b\bar{z}) \mid z \in E\}.$$

The sets $L(a,b), a, b \in E$, are precisely the 2-dimensional subspaces of the $F$-vector space $E^2$ which are complementary to $S = \{0\} \times E$.

Let now $\chi : E^2 \to E^2$ be bijective and semilinear such that $\chi(S) = S$. Assume moreover that the companion automorphism of $\chi$ can be extended to a ring automorphism $\sigma : E \to E$ which commutes with $^-$. It follows from elementary linear algebra that $\chi$ can be written as follows

$$(z, w) \mapsto (\lambda_{a_1, b_1, \sigma}(z), \lambda_{a_2, b_2, \sigma}(z) + \lambda_{a_3, b_3, \sigma}(w))$$

where $a_i, b_i \in E, i = 1, \ldots, 3$, and $N(a_1) - N(b_1) \neq 0 \neq N(a_3) - N(b_3)$. With this notation we have

LEMMA 4.3. *Let $a, b \in E$. Then $\chi(L(a,b)) = L(a', b')$ where*

$$(a', b') = \tfrac{1}{N(a_1)-N(b_1)} \left( a_3(a^\sigma, b^\sigma) + b_3(\bar{b}^\sigma, \bar{a}^\sigma) + (a_2, b_2) \right) \begin{pmatrix} \bar{a}_1 & -b_1 \\ -\bar{b}_1 & a_1 \end{pmatrix}.$$

*Proof.* Let $z \in E$. Then we have

$$\chi(z, \lambda_{a,b}(z)) = (\lambda_{a_1, b_1, \sigma}(z), \lambda_{a_2, b_2, \sigma}(z) + \lambda_{a_3, b_3, \sigma}(\lambda_{a,b}(z))).$$

We put $w = \lambda_{a_1, b_1, \sigma}(z)$. Then we get that $L(a', b')$ consists of all vectors

$$\left( w, (\lambda_{a_2, b_2, \sigma} \circ \lambda_{a_1, b_1, \sigma}^{-1})(w) + (\lambda_{a_3, b_3, \sigma} \circ \lambda_{a,b} \circ \lambda_{a_1, b_1, \sigma}^{-1})(w) \right)$$

where $w \in E$. Using Lemma 4.1 and Lemma 4.2 this yields

$$a' = \tfrac{1}{N(a_1)-N(b_1)} \left( (a_3 a^\sigma + b_3 \bar{b}^\sigma + a_2)\bar{a}_1 - (a_3 b^\sigma + b_3 \bar{a}^\sigma + b_2)\bar{b}_1 \right),$$

$$b' = \tfrac{1}{N(a_1)-N(b_1)} \left( -(a_3 a^\sigma + b_3 \bar{b}^\sigma + a_2)b_1 + (a_3 b^\sigma + b_3 \bar{a}^\sigma + b_2)a_1 \right).$$

Combining these two equations into one matrix equation proves the lemma. $\square$

Let $\mathcal{J} = \{(\varphi(m), \psi(m)) \mid m \in E\} \subset E^2$ be an $E_1$-indicator set and let $\mathcal{B}$ be the corresponding spread. It follows easily from Lemma 4.3 that the pairs of functions $(\psi, \varphi), (\overline{\varphi}, \overline{\psi})$ and $(\overline{\psi}, \overline{\varphi})$ all define $E_1$-indicator sets, whose corresponding spreads are equivalent to $\mathcal{B}$. For the pair $(\varphi, \overline{\psi})$ this is not true in general. Instead we have

LEMMA 4.4. *Let* $\mathcal{J} = \{(\varphi(m), \psi(m)) \mid m \in E\} \subset E^2$ *be an* $E_1$-*indicator set and let* $\mathcal{B}$ *be the corresponding spread. Then* $\mathcal{J}^\circ = \{(\varphi(m), \overline{\psi(m)}) \mid m \in E\}$ *is an* $E_1$-*indicator set if and only if* $\mathcal{B}$ *induces a dual spread* $\mathcal{B}^\circ$. *If this holds true then the spread corresponding to* $\mathcal{J}^\circ$ *is equivalent to* $\mathcal{B}^\circ$.

*Proof.* We define a non-degenerate alternating bilinear form $B : E^2 \times E^2 \to F$ by

$$B((z_1, w_1), (z_2, w_2)) = z_1 w_2 + \overline{z}_1 \overline{w}_2 - w_1 z_2 - \overline{w}_1 \overline{z}_2.$$

Let $a, b, z, w \in E$. Then we have

$$B((z, az + b\overline{z}), (w, aw + \overline{b}\overline{w}))$$
$$= z(aw + \overline{b}\overline{w}) + \overline{z}(\overline{aw} + bw) - (az + b\overline{z})w - (\overline{az} + \overline{b}z)\overline{w} = 0.$$

Since $L(a, b)$ and $L(a, \overline{b})$ are 2-dimensional subspaces of the 4-dimensional vector space $E$, it follows that $L(a, b)^\perp = L(a, \overline{b})$. Moreover, $S^\perp = S$, and hence we can argue exactly as in the proof of Proposition 1.33. $\square$

## 4.3 Indicator Sets and Quasifields

In this section we investigate the quasifield associated with an $E_1$-indicator set.

We use the same notation as in section 4.2. Let $\varphi, \psi : E \to E$ be two arbitrary mappings. Then, $\mathcal{J} = \{(\varphi(m), \psi(m)) \mid m \in E\}$ is an $E_1$-indicator set if and only if $\mathcal{M}(\mathcal{J}) = \{\lambda_{\varphi(m), \psi(m)} \mid m \in E\}$ is a spread set of the 2-dimensional $F$-vector space $E$, cf. Definition 1.10. Recall that a spread set is normalized if it contains the zero and the identity endomorphism. Accordingly, we call an $E_1$-indicator set $\mathcal{J}$ normalized if the spread set $\mathcal{M}(\mathcal{J})$ is normalized, i.e. if $\{(0, 0), (1, 0)\} \subseteq \mathcal{J}$. By Lemma 1.15, every spread set is equivalent to a normalized one.

PROPOSITION 4.5. *Let* $\mathcal{J} = \{(\varphi(m), \psi(m)) \mid m \in E\}$ *be a normalized* $E_1$-*indicator set of* $E^2$. *Put* $\varphi' = \varphi \circ (\varphi + \psi)^{-1}$ *and* $\psi' = \psi \circ (\varphi + \psi)^{-1}$ *and define a multiplication* $* : E \times E \to E$ *by*

$$m * z = \varphi'(m)z + \psi'(m)\overline{z}.$$

*Then* $Q(\mathcal{J}) = (E, +, *)$ *is a left quasifield which coordinatizes the translation plane associated with* $\mathcal{J}$.

*Proof.* We have $\mathcal{J} = \{(\varphi'(m), \psi'(m)) \mid m \in E\}$ and hence we may assume $\varphi' = \varphi$ and $\psi' = \psi$. It follows that $\varphi(m) + \psi(m) = m$ for all $m \in E$. Since $\mathcal{J}$ is normalized, there exist elements $m_0, m_1 \in E$ such that $\varphi(m_0) = \psi(m_0) = \psi(m_1) = 0$ and $\varphi(m_1) = 1$. This yields $m_0 = \varphi(m_0) + \psi(m_0) = 0$ and $m_1 = \varphi(m_1) + \psi(m_1) = 1$.

Let $z \in E$. Then we have

$$z * 0 = \varphi(z)0 + \psi(z)\overline{0} = 0,$$
$$0 * z = \varphi(0)z + \psi(0)\overline{z} = 0,$$
$$z * 1 = \varphi(z)1 + \psi(z)\overline{1} = z,$$
$$1 * z = \varphi(1)z + \psi(1)\overline{z} = z.$$

Consequently, axioms (Q2) and (Q3) are satisfied.

Since $(E, +)$ is an abelian group, (Q1) is certainly satisfied.

Since the elements of $\mathcal{M}(\mathcal{J})$ are linear mappings of the $F$-vector space $E$, axiom (Q4) is satisfied as well. Note that we are dealing with left quasifields, hence the factors in the formulation of (Q4) have to be exchanged.

Axioms (Q5) and (Q6) are satisfied since they are equivalent to (I1) and (I2) or (A1) and (A2), respectively.

It follows that $(E, +, *)$ is a quasifield. The spread associated with this quasifield consists of $S$ and of all sets $\{(z, m * z) \mid z \in E\}, m \in E$. So, the spreads associated with $(E, +, *)$ and $\mathcal{J}$ are the same. $\qquad\square$

PROPOSITION 4.6. *Let* $\mathcal{J} = \{(\varphi(m), \psi(m)) \mid m \in E\}$ *be an* $E_1$-*indicator set with* $(0, 0) \in \mathcal{J}$. *Denote the translation plane associated with* $\mathcal{J}$ *by* $\mathcal{A}$. *Then the following conditions are equivalent:*

(a) *The mappings* $\varphi \circ (\varphi + \psi)^{-1}$ *and* $\psi \circ (\varphi + \psi)^{-1}$ *are additive homomorphisms of* $E$.

(b) *There exists a bijection* $\varrho : E \to E$ *such that* $\varphi \circ \varrho$ *and* $\psi \circ \varrho$ *are additive homomorphisms of* $E$.

(c) *The shear group* $\Sigma_{[s,S]}$ *of* $\mathcal{A}$ *is linearly transitive.*

*Proof.* Obviously, (a) implies (b).

Assume now that there exists $\varrho : E \to E$ such that $\varphi \circ \varrho$ and $\psi \circ \varrho$ are additive homomorphisms of $E$. We have $\mathcal{J} = \{((\varphi \circ \varrho)(m), (\psi \circ \varrho)(m)) \mid m \in E\}$. Hence we may assume that $\varphi$ and $\psi$ are themselves additive. Consider now the mappings

$$\chi_t : E^2 \to E^2 : (z, w) \mapsto (z, \lambda_{\varphi(t), \psi(t)}(z) + w) \quad \text{with } t \in E.$$

These mappings fix all points in $S$ and all 2-dimensional affine subspaces of $E^2$ which are parallel to $S$. Since $\varphi$ and $\psi$ are additive, we get $\chi_t\left(L(\varphi(m), \psi(m))\right) = L(\varphi(m + t), \psi(m + t))$ for all $m, t \in E$. It follows that $\Sigma_{[s,S]} = \{\chi_t \mid t \in E\}$ is a transitive group of shears.

Assume now that $\Sigma_{[s,S]}$ is a transitive group of shears. Then it follows from Proposition 1.20 that $\Sigma_{[s,S]}$ is isomorphic to the additive group of $E$. Since $\Sigma_{[s,S]}$

fixes all points in $S$ and all 2-dimensional affine subspaces parallel to $S$, there are additive homomorphisms $\varphi' : E \to E$ and $\psi' : E \to E$ such that

$$\Sigma_{[s,S]} = \{\chi_t : E^2 \to E^2 : (z,w) \mapsto (z, \lambda_{\varphi'(t),\psi'(t)}(z) + w) \,|\, t \in E\}.$$

Since $L(0,0) \in \mathcal{B}$ and $\Sigma_{[s,S]}$ operates transitively on $\mathcal{B} \setminus \{S\}$, we get $\mathcal{B} \setminus \{S\} = \{L(\varphi'(t), \psi'(t)) \,|\, t \in E\}$. Hence there exists a bijection $\varrho : E \to E$ such that $\varphi' = \varphi \circ \varrho$ and $\psi' = \psi \circ \varrho$. We have

$$\begin{aligned}
\varphi' \circ (\varphi' + \psi')^{-1} &= \varphi \circ \varrho \circ (\varphi \circ \varrho + \psi \circ \varrho)^{-1} \\
&= \varphi \circ \varrho \circ ((\varphi + \psi) \circ \varrho)^{-1} \\
&= \varphi \circ \varrho \circ \varrho^{-1} \circ (\varphi + \psi)^{-1} \\
&= \varphi \circ (\varphi + \psi)^{-1}.
\end{aligned}$$

It follows that $\varphi \circ (\varphi + \psi)^{-1}$ is an additive homomorphism of $E$. By the same argument, $\psi \circ (\varphi + \psi)^{-1}$ is also an additive homomorphism, and hence (a) is satisfied. $\qquad\square$

COROLLARY 4.7. *Let* $\mathcal{J} = \{(\varphi(m), \psi(m)) \,|\, m \in E\} \subset E^2$ *be a normalized* $E_1$-*indicator set. Then the quasifield* $Q(\mathcal{J}) = (E, +, *)$ *is a semifield if and only if the mappings* $\varphi \circ (\varphi + \psi)^{-1}$ *and* $\psi \circ (\varphi + \psi)^{-1}$ *are additive homomorphisms of* $E$.

PROPOSITION 4.8. *Let* $\mathcal{J} = \{(\varphi(m), \psi(m)) \,|\, m \in E\} \subset E^2$ *be a normalized* $E_1$-*indicator set. Then the following conditions are equivalent:*

(a) *The mappings* $\varphi \circ (\varphi + \psi)^{-1}$ *and* $\psi \circ (\varphi + \psi)^{-1}$ *are linear mappings of the* $F$-*vector space* $E$.

(b) *There exists a bijection* $\varrho : E \to E$ *such that* $\varphi \circ \varrho$ *and* $\psi \circ \varrho$ *are linear mappings of the* $F$-*vector space* $E$.

(c) *The translation plane* $\mathcal{A}$ *associated with* $\mathcal{J}$ *is pappian.*

*Proof.* Obviously, (a) implies (b).

Let now (b) be satisfied. Then we may assume that $\varphi$ and $\psi$ are themselves linear. Moreover, we may assume that $\varphi(m) + \psi(m) = m$ for all $m \in E$.

It is sufficient to show that the quasifield $Q(\mathcal{J}) = (E, +, *)$ is a commutative field.

Since $\varphi$ is linear, there exist $a, b \in E$ such that $\varphi(x) = ax + b\overline{x}$ for all $x \in E$. As $\mathcal{J}$ is normalized, we have $\varphi(1) = a + b = 1$. From $\varphi + \psi = \mathrm{id}$ we infer that

$$\varphi(x) = (1-b)x + b\overline{x} \quad \text{and} \quad \psi(x) = bx - b\overline{x}$$

for all $x \in E$. Thus we get

$$\begin{aligned}
x * y &= \varphi(x)y + \psi(x)\overline{y} \\
&= (1-b)xy + b\overline{x}y + bx\overline{y} - b\overline{x}\overline{y} \\
&= xy - b(x - \overline{x})(y - \overline{y})
\end{aligned}$$

for $x, y \in E$. This shows that the multiplication $*$ is commutative.

To see that the multiplication is also associative, we proceed as follows. Let $x, y, z \in E$. Then

$$\begin{aligned}
(x * y) * z &= (xy - b(x - \bar{x})(y - \bar{y}))\, z \\
&\quad - b\left(xy - b(x - \bar{x})(y - \bar{y}) - \overline{xy} + \bar{b}(\bar{x} - x)(\bar{y} - y)\right)(z - \bar{z}) \\
&= xyz - b(x - \bar{x})(y - \bar{y})z - bxy(z - \bar{z}) + b^2(x - \bar{x})(y - \bar{y})(z - \bar{z}) \\
&\quad + b\overline{xy}(z - \bar{z}) - b\bar{b}(\bar{x} - x)(\bar{y} - y)(z - \bar{z}) \\
&= xyz + b^2(x - \bar{x})(y - \bar{y})(z - \bar{z}) + b\bar{b}(\bar{x} - x)(\bar{y} - y)(\bar{z} - z) \\
&\quad + b(-2xyz - \overline{xyz} + xy\bar{z} + x\bar{y}z + \bar{x}yz).
\end{aligned}$$

This expression is symmetric in $x, y$ and $z$ and so we have

$$(x * y) * z = (y * z) * x = x * (y * z)$$

since we already now that $*$ is commutative. It follows that $(E, +, *)$ is a commutative field and hence $\mathcal{A}$ is pappian.

Assume now that $\mathcal{A}$ is pappian. Put $\varphi' = \varphi \circ (\varphi + \psi)^{-1}$ and $\psi' = \psi \circ (\varphi + \psi)^{-1}$. Then the quasifield $(E, +, *)$ is a commutative field, where $x * y = \varphi'(x)y + \psi'(x)\bar{y}$ for $x, y \in E$.

Corollary 4.7 implies that $\varphi'$ and $\psi'$ are additive homomorphism of $E$.

Let $x, z \in E$ and $y \in F$. Then we have

$$x * y = \varphi'(x)y + \psi'(x)\bar{y} = \varphi'(x)y + \psi'(x)y = xy$$

since $\bar{y} = y$ and $\varphi(x) + \psi(x) = x$. As $*$ is commutative, we also get $y * z = yz$. It follows that

$$\begin{aligned}
\varphi'(xy)z + \psi'(xy)\bar{z} &= (xy) * z = (x * y) * z \\
&= x * (y * z) = x * (yz) = \varphi'(x)yz + \psi'(x)\overline{yz}.
\end{aligned}$$

Since $\varphi' + \psi' = \mathrm{id}$ and $\bar{y} = y$ this yields

$$\varphi'(xy)z + xy\bar{z} - \varphi'(xy)\bar{z} = \varphi'(x)yz + xy\bar{z} - \varphi'(x)y\bar{z}$$

and hence

$$\varphi'(xy)(z - \bar{z}) = \varphi'(x)y(z - \bar{z}).$$

We can choose $z$ such that $z - \bar{z}$ is invertible and hence $\varphi'$ is $F$-linear. $\qquad\square$

If we drop the assumption that $\mathcal{J}$ is normalized the proposition does not hold any longer. This follows from Lemma 4.3 if we take for $\sigma$ a proper field automorphism.

Let $(E, +, \cdot)$ be a separable quadratic extension field of $F$. It follows from the proof of Proposition 4.8 that all other multiplications $* : E \times E \to E$ which turn $(E, +, *)$ into a quadratic extension field of $F$ such the multiplicative identities of $\cdot$ and $*$ coincide, are obtained as follows. There exists an element $b \in E$ such that $x * y = x \cdot y - b \cdot (x - \bar{x}) \cdot (y - \bar{y})$ for $x, y \in E$.

There seems to be no simple characterization of the $E_1$-indicator sets which lead to desarguesian planes, cf. the remarks on the quaternion plane after Proposition 6.5.


## 4.4 Reguli and Flocks of Reguli

The investigations in this section are inspired by the paper [80] of R. Metz.

DEFINITION 4.9. Let $\mathcal{P} = (P, \mathcal{L})$ be a projective space over a skewfield $F$. A system $\mathcal{R}$ of subspaces of $\mathcal{P}$ with $|\mathcal{R}| > 2$ is called a *regulus* of $\mathcal{P}$ if the following conditions are satisfied:

(R1) Any two different elements of $\mathcal{R}$ are complementary, i.e. $\mathcal{R}$ is a partial spread.

(R2) Each line of $\mathcal{P}$ which intersects 3 different elements of $\mathcal{R}$ intersects every element of $\mathcal{R}$. Such a line is called a *transversal* of $\mathcal{R}$.

(R3) Each point of $\mathcal{P}$ which is contained in a transversal of $\mathcal{R}$ is also contained in an element of $\mathcal{R}$.

If a finite-dimensional projective space contains 3 mutually complementary subspaces its dimension is necessarily an odd number. The *generalized theorem of Dandelin* [56; 36: Proposition] states that if a projective space $\mathcal{P}$ of dimension at least 3 contains a regulus then $\mathcal{P}$ is pappian. Moreover, if $X, Y, Z$ are mutually complementary subspaces of a pappian projective space $\mathcal{P}$ then there exists precisely one regulus $\mathcal{R}(X, Y, Z)$ such that $X, Y, Z \in \mathcal{R}(X, Y, Z)$.

DEFINITION 4.10. Let $\mathcal{P}$ be a projective space over a commutative field $F$ and let $\mathcal{B}$ be a spread of $\mathcal{P}$.

(a) Let $S, W \in \mathcal{B}, S \neq W$. The spread $\mathcal{B}$ is called $(S, W)$-*regular* if $\mathcal{B}$ contains all reguli $\mathcal{R}(S, W, U)$ where $U \in \mathcal{B} \setminus \{S, W\}$.

(b) Let $S \in \mathcal{B}$. The spread $\mathcal{B}$ is called $S$-*regular* if it is $(S, W)$-regular for all $W \in \mathcal{B} \setminus \{S\}$.

(c) The spread $\mathcal{B}$ is called *regular* if it is $S$-regular for all $S \in \mathcal{B}$.

It should be noted that the notion of regularity depends not only the spread $\mathcal{B}$ but also on the field $F$. Assume for example that $V = \mathbb{H}^2$ and let $\mathcal{B}$ consist of all 1-dimensional $\mathbb{H}$-subspaces of $V$. Then, $\mathcal{B}$ is regular if we think of $V$ as an 8-dimensional vector space over $\mathbb{R}$. However, if we view $V$ as a 4-dimensional vector space over $\mathbb{C}$ then $\mathcal{B}$ is not regular.

In a projective space over the field with 2 elements every spread is regular. Let now $\mathcal{P}$ be a projective space over a commutative field $F$ with $|F| > 2$. Furthermore, let $\mathcal{B}$ be a spread of $\mathcal{P}$ and let $\mathcal{A}$ denote the corresponding translation plane. Then we have:

(a) If $\mathcal{B}$ is $S$-regular then $\Sigma_{[s,S]}$ is a transitive shear group of $\mathcal{A}$ [75: Teorema 5; 80: Satz 4.5].

(b) If $\mathcal{B}$ is regular then $\mathcal{A}$ is a Moufang plane [19: Theorem 12.1].

If $\mathcal{P}$ is 3-dimensional this result can be sharpened, cf. Proposition 4.13.

From now we assume that $\mathcal{P}$ is 3-dimensional and we use the notation from section 4.2. Our first result is

LEMMA 4.11. *Let* $a_1, a_2, b_1, b_2 \in E$ *with* $N(a_1 - a_2) \neq N(b_1 - b_2)$. *Then we have*

$$\mathcal{R}(S, L(a_1, b_1), L(a_2, b_2)) = \{S\} \cup \{L((a_1 - a_2)t + a_2, (b_1 - b_2)t + b_2) \mid t \in F\}.$$

*Proof.* Assume first that $a_1 = 1$ and $a_2 = b_1 = b_2 = 0$. In this case, the lemma asserts that the regulus spanned by $S, L(0,0) = E \times \{0\}$ and $L(1,0) = \{(z,z) \mid z \in E\}$ consists of $S$ and of the sets $L(t,0) = \{(z, tz) \mid z \in E\}$ with $t \in F$. But this is a well known fact. Consider now the linear mapping

$$\chi : E^2 \to E^2 : (z, w) \mapsto (z, \lambda_{a_2, b_2}(z) + \lambda_{a_1 - a_2, b_1 - b_2}(w)).$$

We have $\chi(S) = S, \chi(L(1,0)) = L(a_1, b_1)$ and $\chi(L(0,0)) = L(a_2, b_2)$. Since the image of a regulus under a linear bijection is again a regulus, we conclude that $\chi(\mathcal{R}(S, L(1,0), L(0,0))) = \mathcal{R}(S, L(a_1, b_1), L(a_2, b_2))$. It follows from Lemma 4.3 that

$$\chi(L(t,0)) = L((a_1 - a_2)t + a_2, (b_1 - b_2)t + b_2)$$

for $t \in F$. This completes the proof. $\qquad\square$

If we think of $E^2$ as a 4-dimensional affine space over $F$, then Lemma 4.11 says that each lines of this affine space whose intersection point with the projective space at infinity is not contained in a certain quadric corresponds to a regulus of $\mathcal{P}$ which contain $S$. This is the point of view taken by R. Metz in [80]. He also gives a description of the lines which do not correspond to reguli and of the higher dimensional subspaces of the affine space $E^2$ [80: 3].

Let $a_1, a_2, b_1, b_2 \in E$ with $N(a_1) \neq N(b_1)$. We denote the regulus $\{S\} \cup \{L(a_1 t + a_2, b_1 t + b_2) \mid t \in F\}$ by $\mathcal{R}(a_1, a_2, b_1, b_2)$.

The following lemma is fundamental for our investigations:

LEMMA 4.12. *Let* $\mathcal{B}$ *be a spread of* $E^2$ *with* $S \in \mathcal{B}$. *Assume that* $\mathcal{B}$ *contains two different reguli* $\mathcal{R}(a_1, a_2, b_1, b_2)$ *and* $\mathcal{R}(a_1', a_2', b_1', b_2')$. *Then the lines* $\{(a_1 t + a_2, b_1 t + b_2) \mid t \in F\}$ *and* $\{(a_1' t + a_2', b_1' t + b_2') \mid t \in F\}$ *in the 4-dimensional affine space* $E^2$ *are either parallel or they intersect.*

*Proof.* Let $\mathcal{J} \subseteq E^2$ be the $E_1$-indicator set associated with $\mathcal{B}$. Put $G_1 = \{(a_1 t + a_2, b_1 t + b_2) \mid t \in F\}$ and $G_2 = \{(a_1' s + a_2', b_1' s + b_2') \mid s \in F\}$. Assume that $G_1 \cap G_2 = \emptyset$. Since $G_1 \cup G_2 \subseteq \mathcal{J}$, we have $N(a_1 t - a_1' s + a_2 - a_2') \neq N(b_1 t -$

$b_1's + b_2 - b_2')$ for all $s, t \in F$. This can be rephrased as follows. The affine subspace $D = \{(a_1t - a_1's + a_2 - a_2, b_1t - b_1's + b_2 - b_2') \mid t, s \in F\} \subseteq E^2$ does not intersect the quadric $\{(z, w) \in E^2 \mid N(z) = N(w)\}$. This can only happen if $D$ is either 1-dimensional or passes through the origin. Since $G_1 \cap G_2 = \emptyset$, the second case is ruled out. So, $D$ is 1-dimensional and hence $G_1$ and $G_2$ are parallel. $\square$

Lemma 4.12 was implicitly proved by Metz in the course of the proof of [80: Satz 4.9]. After appropriate reformulations Lemma 4.12 is also equivalent to [54: (1)] and [33: Lemma 2.1].

PROPOSITION 4.13. *Let $\mathcal{P}$ be a 3-dimensional projective space over a commutative field $F$ and let $\mathcal{B}$ be a spread of $\mathcal{P}$. Then the following conditions are equivalent:*

(a) *The translation plane associated with $\mathcal{B}$ is pappian.*

(b) *$\mathcal{B}$ is regular.*

(c) *There exists a line $S \in \mathcal{B}$ such that $\mathcal{B}$ is $S$-regular.*

(d) *There are 2 different lines $S, W \in \mathcal{B}$ such that $\mathcal{B}$ is $(S, W)$-regular. Moreover, $\mathcal{B}$ contains a regulus which contains $S$ but not $W$.*

*Proof.* Each affine plane of order 4 is pappian, hence we may assume $|F| \geq 3$.

It is known that (a) implies (b) [19: Theorem 12.1] and obviously (b) implies (c).

Assume that (c) is satisfied. Let $\mathcal{J} \subseteq E^2$ be the $E_1$-indicator set corresponding to $\mathcal{B}$. From Lemma 4.11 we infer that $\mathcal{J}$ is a subspace of the 4-dimensional affine space $E^2$. But then (d) also holds.

It remains to show that (a) follows from (d). Let $\mathcal{J} \subseteq E^2$ be the $E_1$-indicator set associated with $\mathcal{B}$. Up to isomorphism, we may assume that $W = L(0, 0)$ and that $(1, 0) \in \mathcal{J}$. This implies in particular that $\mathcal{J}$ is normalized. From Lemma 4.11 we infer that $\mathcal{J}$ is a union of lines of the affine space $E^2$ all of which contain the point $(0, 0)$. Let $\mathcal{R}(a_1, a_2, b_1, b_2) \subset \mathcal{B}$ be a regulus which contains $S$ but not $W$ and let $G$ be a line of the affine space $E^2$ with $G \subseteq \mathcal{J}$. By Lemma 4.12, the lines $\{(a_1t + a_2, b_1t + b_2) \mid t \in F\}$ and $G$ either are parallel or they intersect non-trivially. But this can only happen if $\mathcal{J}$ is an affine plane in $E^2$. Since $(0, 0) \in \mathcal{J}$, it follows that there exist $F$-linear mappings $\varphi, \psi : E \to E$ such that $\mathcal{J} = \{(\varphi(m), \psi(m)) \mid m \in E\}$. By Proposition 4.8, the translation plane $\mathcal{A}$ is pappian. $\square$

We now prove the main result of this section.

THEOREM 4.14. *Let $\mathcal{B}$ be a spread of $E^2$ with $S \in \mathcal{B}$. Assume that for each $U \in \mathcal{B} \setminus \{S\}$ there exists a regulus $\mathcal{R}$ with $U, S \in \mathcal{R}$ and $\mathcal{R} \subset \mathcal{B}$. Then precisely one of the following three cases is realized:*

(a) *The translation plane associated with $\mathcal{B}$ is pappian.*

(b) *For each $U \in \mathcal{B} \setminus \{S\}$ there is precisely one regulus $\mathcal{R}$ such that $U, S \in \mathcal{R}$ and $\mathcal{R} \subset \mathcal{B}$.*

(c) *There is a subspace $W \in \mathcal{B} \setminus \{S\}$ such that each regulus $\mathcal{R}$ with $S \in \mathcal{R}$ and $\mathcal{R} \subset \mathcal{B}$ is of the form $\mathcal{R}(S, W, U)$ where $U \in \mathcal{B} \setminus \{S, W\}$.*

*Proof.* Let $\mathcal{J} \subseteq E^2$ be the $E_1$-indicator set associated with $\mathcal{B}$. Lemma 4.12 implies that $\mathcal{J}$ is covered by lines of the 4-dimensional affine space $E^2$ any two of which are either parallel or intersect. If both possibilities occur, then $\mathcal{J}$ is contained in a plane of the affine space $E^2$. Hence there are elements $a_i, b_i \in E, i = 0, \ldots, 2$, such that for each $m \in E$ there is exactly one pair $(x, y) \in F^2$ with

$$(\varphi(m), \psi(m)) = (a_0, b_0) + (a_1, b_1)x + (a_2, b_2)y.$$

Since $\varphi - \psi$ is bijective, it follows that $x$ and $y$ take on all values in $F$. Hence, $\mathcal{J}$ is itself a plane of the affine space $E^2$. From Lemma 4.11 we conclude that $\mathcal{B}$ is $S$-regular and by Proposition 4.13 we are in case (a).

If $\mathcal{J}$ is not an affine plane in $E^2$, then either all affine lines in $\mathcal{J}$ are parallel or they all pass through the same point. Hence, we are either in case (b) or in case (c). $\square$

COROLLARY 4.15. *Let $\mathcal{B}$ be spread of a 3-dimensional pappian projective space $\mathcal{P}$. Then the translation plane associated with $\mathcal{B}$ is pappian if and only if for each $Z, W \in \mathcal{B}$ there exists a regulus $\mathcal{R}$ of $\mathcal{P}$ with $Z, W \in \mathcal{R}$ and $\mathcal{R} \subset \mathcal{B}$.*

Theorem 4.14 motivates the following definition.

DEFINITION 4.16. Let $\mathcal{P}$ be a 3-dimensional projective space over a commutative field $F$ and let $\mathcal{B}$ be a spread of $\mathcal{P}$. Furthermore, let $\mathcal{Y}$ be a system of reguli contained in $\mathcal{B}$.

(a) Let $S \in \mathcal{B}$. Then $\mathcal{Y}$ is called a *parabolic flock of reguli with carrier* $S$ if $S$ is contained in every regulus of $\mathcal{Y}$ and if each $U \in \mathcal{B} \setminus \{S\}$ is contained in exactly one regulus of $\mathcal{Y}$.

(b) Let $S, W \in \mathcal{B}$ with $S \neq W$. Then $\mathcal{Y}$ is called a *hyperbolic flock of reguli with carrier* $(S, W)$ if $S$ and $W$ are contained in every regulus of $\mathcal{Y}$ and if each $U \in \mathcal{B} \setminus \{S, W\}$ is contained in exactly one regulus of $\mathcal{Y}$.

Obviously, a spread of a 3-dimensional projective space contains a hyperbolic flock of reguli with carrier $(S, W)$ if and only it is $(S, W)$-regular.

PROPOSITION 4.17. *Let $\mathcal{P}$ be a 3-dimensional projective space over a commutative field $F$ and let $\mathcal{B}$ be a spread of $\mathcal{P}$.*

(a) *Let $S \in \mathcal{B}$. The spread $\mathcal{B}$ contains a parabolic flock of reguli with carrier $S$ if and only if there exists a group $\Delta \leq \Sigma_{[s,S]}$ and an element $U \in \mathcal{B} \setminus \{S\}$ such that $\{S\} \cup \{\chi(U) \mid \chi \in \Delta\}$ is a regulus of $\mathcal{P}$.*

*(b) Let $S, W \in \mathcal{B}$ with $S \neq W$. The spread $\mathcal{B}$ contains a hyperbolic flock of reguli with carrier $(S, W)$ if and only if there exists a group $\Delta \leq \Sigma_{[w,S]}$ and an element $U \in \mathcal{B} \setminus \{S, W\}$ such that $\{S, W\} \cup \{\chi(U) \,|\, \chi \in \Delta\}$ is a regulus of $\mathcal{P}$.*

*Proof.* We think of $\mathcal{P}$ as projective space over the 4-dimensional $F$-vector space $A^2$. Furthermore, we put $S = \{0\} \times A$ and we describe $\mathcal{B}$ using an $A_1$-indicator set $\mathcal{J}$, cf. section 2.7.

(a) Let $S$ be the carrier of a parabolic flock of reguli. Lemma 4.12 implies that $\mathcal{J}$ is the disjoint union of mutually parallel lines of the 4-dimensional affine space $A^2$. Up to isomorphism, we may assume that if $(a, b) \in \mathcal{J}$ then also $(a + t, b) \in \mathcal{J}$ for all $t \in F$. From Lemma 4.3 we infer that the set of all mappings

$$A^2 \to A^2 : (z, w) \mapsto (z, tz + w) \quad \text{with } t \in F$$

is a group as required.

Let now $\mathcal{R} \subset \mathcal{B}$ be a regulus with $S \in \mathcal{R}$ and assume that $\mathcal{R} \setminus \{S\}$ is an orbit of a subgroup $\Delta$ of $\Sigma_{[s,S]}$. Up to isomorphism, we may assume $\mathcal{R} \setminus \{S\} = \{L(t, 0) \,|\, t \in F\}$. All elements of $\Sigma_{[s,S]}$ are of the form

$$A^2 \to A^2 : (z, w) \mapsto (z, \lambda_{a,b}(z) + w) \quad \text{where } a, b \in A.$$

As $\mathcal{R} \setminus \{S\}$ is an orbit of $\Delta$, it follows from Lemma 4.3 that $\Delta$ consists of all these mappings with $a \in F$ and $b = 0$. The unions of the other orbits of $\Delta$ in $\mathcal{B}$ with $S$ are then reguli as well. Thus, we have found a parabolic flock of reguli in $\mathcal{B}$.

(b) Let $W \in \mathcal{B} \setminus \{S\}$ and let $(S, W)$ be the carrier of a hyperbolic flock of reguli. Up to isomorphism, we may assume $W = L(0, 0)$. Lemma 4.12 implies that $\mathcal{J}$ is the union of 1-dimensional subspaces of the 4-dimensional vector space $A^2$. From Lemma 4.3 we infer that the set of all mappings

$$A^2 \to A^2 : (z, w) \mapsto (z, sw) \quad \text{with } s \in F^\times$$

is a group as required.

Let now $W \in \mathcal{B} \setminus \{S\}$ and let $\mathcal{R}$ be a regulus with $S, W \in \mathcal{R}$ and $\mathcal{R} \subset \mathcal{B}$. Assume further that $\mathcal{R} \setminus \{S, W\}$ is an orbit of a subgroup $\Delta$ of $\Sigma_{[w,S]}$. Up to isomorphism, we may assume $\mathcal{R} \setminus \{S\} = \{L(t, 0) \,|\, t \in F\}$ and $W = L(0, 0)$. All elements of $\Sigma_{[w,S]}$ are of the form

$$A^2 \to A^2 : (z, w) \mapsto (z, \lambda_{a,b}(w)) \quad \text{where } a, b \in A \text{ and } N(a) \neq N(b).$$

As $\mathcal{R} \setminus \{S, W\}$ is an orbit of $\Delta$, it follows from Lemma 4.3 that $\Delta$ consists of all these mappings with $a \in F^\times$ and $b = 0$. The unions of the other orbits of $\Delta$ in $\mathcal{B}$ with $S$ and $W$ are then reguli as well. Thus, we have found a hyperbolic flock of reguli in $\mathcal{B}$. $\qquad\square$

For finite projective spaces, (a) was first proved by Johnson [64: Theorem A] and (b) by Gevaert and Johnson [32: Theorem 3.4].

We now make some remarks on the connection between flocks of reguli in spreads and flocks of Miquelian Benz planes. For this, we need some preparations.

For a proof of the following lemma the reader is referred to Brauner [15: 10.2].

LEMMA 4.18. *Let $P$ be a 3-dimensional projective space over a commutative field $F$.*

(a) *The transversals of a regulus $\mathcal{R}$ of $P$ form themselves a regulus $\mathcal{R}'$ and we have $\mathcal{R}'' = \mathcal{R}$. The regulus $\mathcal{R}'$ is called the opposite regulus of $\mathcal{R}$.*

(b) *Let $S$ and $W$ be two non-intersecting lines of $P$. Furthermore, let $\pi : S \to W$ be a projective mapping. Then $\mathcal{R} = \{p \vee \pi(p) \mid p \in S\}$ is a regulus of $P$. Every regulus of $P$ which admits $S$ and $W$ as transversals can be obtained in this way.*

Let $P$ be a 3-dimensional projective space over a commutative field $F$ and let $S$ and $W$ be two non-intersecting lines of $P$. We define a circle geometry $\mathcal{M}$ as follows: Points of $\mathcal{M}$ are all lines of $P$ which intersect $S$ and $W$. Circles of $\mathcal{M}$ are all reguli of $P$ which admit $S$ and $W$ as transversals. The incidence is given by containment. It follows from Lemma 4.18 (b) and [5: III.§4.Satz 5.1] that $\mathcal{M}$ is isomorphic to the Miquelian Minkowski plane over $F$.

Let now $\mathcal{Y}$ be a hyperbolic flock of reguli with carrier $(S, W)$. We replace all reguli contained in $\mathcal{Y}$ by their opposite reguli. The set $\mathcal{F}$ of circles of $\mathcal{M}$ thus obtained has the following property. Every point of $\mathcal{M}$ is contained in precisely one element of $\mathcal{F}$. According to [31], we call $\mathcal{F}$ a *flock* of $\mathcal{M}$.

The construction just described can be reversed. Let $\mathcal{F}$ be a flock of $\mathcal{M}$. Then $\mathcal{Y} = \{\mathcal{R}' \mid \mathcal{R} \in \mathcal{F}\}$ is a hyperbolic flock of reguli covering a spread $\mathcal{B}$ of $P$. This fact was first noted independently by Thas and Walker [31].

All flocks of finite Miquelian Minkowski planes were recently classified by Bader and Lunardon [4]. The corresponding translation planes turned out to be nearfield planes.

There is an analogous correspondence between flocks of Miquelian Laguerre planes and parabolic flocks of reguli. This was proved by Gevaert, Johnson and Thas [33] for the case of finite fields and by Heimbeck [54] in full generality.

Fisher and Thas [31] describe a method which allows one to associate a spread with each flock of a Miquelian inversive plane. Each spread $\mathcal{B}$ obtained in this way contains two different lines $S, W$ such that $\mathcal{B} \setminus \{S, W\}$ is covered by mutually disjoint reguli. Spreads having this property were studied by Sherk and Pabst in [98] under the name of md-spreads. They also give some examples of md-spreads which are not associated with flocks of Miquelian inversive planes.

# 5. Locally Compact 4-dimensional Translation Planes

In this chapter we study topological spreads of a 4-dimensional real vector space and their description using $\mathbb{C}_1$-indicator sets.

## 5.1 Topological Spreads of a 4-dimensional Real Vector Space

We use the notation of section 4.2 with $F = \mathbb{R}$ and $E = \mathbb{C}$. In particular, $L(a, b)$ denotes the graph of the linear mapping $\lambda_{a,b} : \mathbb{C} \to \mathbb{C} : z \mapsto az + b\bar{z}$ for $a, b \in \mathbb{C}$.

THEOREM 5.1. *Let $B$ be a topological spread of the 4-dimensional real vector space $\mathbb{C}^2$ such that $S = \{0\} \times \mathbb{C} \in B$. Then there exists a mapping $\varphi : \mathbb{C} \to \mathbb{C}$ which satisfies the following conditions:*

> *(K1) The mapping $\varphi$ is a contraction, i.e. for all $m, n \in \mathbb{C}$ with $m \neq n$ we have $|\varphi(m) - \varphi(n)| < |m - n|$.*
>
> *(K2) $\lim\limits_{|m| \to \infty} ||m| - |\varphi(m)|| = \infty$.*

> *$B \setminus \{S\}$ consists either of all subspaces $L(m, \varphi(m)), m \in \mathbb{C}$, or of all subspaces $L(\varphi(m), m), m \in \mathbb{C}$.*
> *Conversely, if $\varphi : \mathbb{C} \to \mathbb{C}$ satisfies (K1) and (K2) then $B = \{S\} \cup \{L(m, \varphi(m)) \mid m \in \mathbb{C}\}$ and $B' = \{S\} \cup \{L(\varphi(m), m) \mid m \in \mathbb{C}\}$ are topological spreads of $\mathbb{C}^2$. These two spreads are equivalent.*

*Proof.* Assume first that $B$ is a topological spread of $\mathbb{C}^2$ which contains $S$. As in Theorem 2.6, we describe $B$ using a $\mathbb{C}_1$-indicator set $\mathcal{J} = \{(\varphi(m), \psi(m)) \mid m \in \mathbb{C}\} \subset \mathbb{C}^2$. The functions $\varphi, \psi : \mathbb{C} \to \mathbb{C}$ satisfy the conditions (C1) and (C2) from Proposition 2.5 and $\Phi : \mathbb{C} \to \mathbb{C} \times \mathbb{C} : m \mapsto (\varphi(m), \psi(m))$ is injective. We have $B = \{S\} \cup \{L(\varphi(m), \psi(m)) \mid m \in \mathbb{C}\}$. From Proposition 1.27 we infer that $B \setminus \{S\}$ and hence also $\mathcal{J}$ are homeomorphic to $\mathbb{C}$. Thus we may assume that $\varphi$ and $\psi$ are continuous. The absolute value of a complex number $z$ is the square root of the norm of $z$. Two complex numbers have the same absolute value if and only if they have the same norm. So we may replace the norm by the absolute value in the formulation of (C1). For $m \neq n$ we have $|\varphi(m) - \varphi(n)| \neq |\psi(m) - \psi(n)|$. The set $\{(m, n) \in \mathbb{C}^2 \,|\, m \neq n\} \subset \mathbb{C}^2$ is connected and $\varphi$ and $\psi$ are continuous. It follows that for $m \neq n$ we always have either

$(I)$ $$|\varphi(m) - \varphi(n)| < |\psi(m) - \psi(n)|$$

or

$(II)$ $$|\varphi(m) - \varphi(n)| > |\psi(m) - \psi(n)|.$$

Assume first that (I) is satisfied. Then $\psi$ is injective. We claim that $\psi$ is also surjective. By Proposition 1.25 it is sufficient to show that $\psi$ is proper. Assume that this is not true. Then we can find a divergent sequence $(m_k)$ in $\mathbb{C}$ such that $|\psi(m_k)|$ is bounded. Since (I) is satisfied $|\varphi(m_k)|$ is bounded as well. Condition (C2) implies that $\varphi + \psi$ is bijective and hence proper, leading to a contradiction. So $\psi$ is bijective and hence we may assume that $\psi$ is the identity. Then (I) becomes (K1).

In order to prove (K2) we use the Klein quadric, cf. section 2.1. We take coordinates from the 6-dimensional real vector space $\mathbb{C}^2 \times \mathbb{R}^2$. The Pfaffian is given by $\mathrm{Pf}(a, b, x, y) = a\bar{a} - b\bar{b} - xy$ for $a, b \in \mathbb{C}, x, y \in \mathbb{R}$. With the line $L(a, b)$ we associate the point $\langle(a, b, 1, a\bar{a} - b\bar{b})\rangle$. The Plücker coordinates of $S$ are given by $\langle(0, 0, 0, 1)\rangle$. The image of $\mathcal{B}$ under the Plücker mapping consists of $\langle(0, 0, 0, 1)\rangle$ and of all points $\langle(\varphi(m), m, 1, |\varphi(m)|^2 - |m|^2)\rangle, m \in \mathbb{C}$. The set of these points is homeomorphic to $\mathbb{S}_2$. Let $(m_k)$ be a sequence of complex numbers with $\lim_{k \to \infty} |m_k| = \infty$. Then we have

$$\lim_{k \to \infty} \langle(\varphi(m_k), m_k, 1, |\varphi(m_k)|^2 - |m_k|^2)\rangle = \langle(0, 0, 0, 1)\rangle.$$

Dividing the left hand side of this equation by $|\varphi(m_k)|^2 - |m_k|^2$ gives us

$$\lim_{k \to \infty} \frac{|\varphi(m_k)|}{|\varphi(m_k)|^2 - |m_k|^2} = \lim_{k \to \infty} \frac{|m_k|}{|\varphi(m_k)|^2 - |m_k|^2} = 0.$$

We add these two equations and get

$$\lim_{k \to \infty} \frac{|\varphi(m_k)| + |m_k|}{|\varphi(m_k)|^2 - |m_k|^2} = \lim_{k \to \infty} \frac{1}{|\varphi(m_k)| - |m_k|} = 0.$$

So we have

$$\lim_{k \to \infty} ||\varphi(m_k)| - |m_k|| = \infty,$$

giving us (K2).

Case (II) is completely analogous.

Let now $\varphi : \mathbb{C} \to \mathbb{C}$ be a mapping which satisfies (K1) and (K2). We show that $\mathcal{B}' = \{S\} \cup \{L(\varphi(m), m) \,|\, m \in \mathbb{C}\}$ is a topological spread of $\mathbb{C}^2$.

First, (K1) implies that the elements of $\mathcal{B}'$ are mutually complementary. If we can show that $\mathcal{B}'$ is homeomorphic to $\mathbb{S}_2$ then we can apply Proposition 1.27 to complete the proof. To this end we embed $\mathcal{B}'$ into the Klein quadric as above. The mapping $\varphi$ is continuous by (K1) and hence the mapping $m \mapsto$

$\langle(\varphi(m), m, 1, |\varphi(m)|^2 - |m|^2)\rangle$ is continuous as well. Moreover, this mapping is injective. Therefore it is sufficient to show that for each sequence $(m_k)$ in $\mathbb{C}$ with $\lim_{k\to\infty} |m_k| = \infty$ the sequence $\langle(\varphi(m_k), m_k, 1, |\varphi(m_k)|^2 - |m_k|^2)\rangle$ converges to $\langle(0, 0, 0, 1)\rangle$. This happens if and only if

$$\lim_{k\to\infty} \frac{|\varphi(m_k)|}{|\varphi(m_k)|^2 - |m_k|^2} = \lim_{k\to\infty} \frac{|m_k|}{|\varphi(m_k)|^2 - |m_k|^2} = \lim_{k\to\infty} \frac{1}{|\varphi(m_k)|^2 - |m_k|^2} = 0.$$

We first compute

$$\lim_{k\to\infty} \frac{|\varphi(m_k)| + |m_k|}{|\varphi(m_k)|^2 - |m_k|^2} = \lim_{k\to\infty} \frac{1}{|\varphi(m_k)| - |m_k|} = 0, \quad \text{(because of (K1))}$$

$$\lim_{k\to\infty} \frac{|\varphi(m_k)| - |m_k|}{|\varphi(m_k)|^2 - |m_k|^2} = \lim_{k\to\infty} \frac{1}{|\varphi(m_k)| + |m_k|} = 0 \quad \text{and}$$

$$\lim_{k\to\infty} \frac{1}{|\varphi(m_k)|^2 - |m_k|^2} = \lim_{k\to\infty} \left( \frac{1}{|\varphi(m_k)| + |m_k|} \cdot \frac{1}{|\varphi(m_k)| - |m_k|} \right) = 0.$$

By adding and subtracting the first and the second of these three equations we get the required result.

By the same reasoning, $\{S\} \cup \{L(m, \varphi(m)) \mid m \in \mathbb{C}\}$ is also a topological spread of $\mathbb{C}^2$.

It is easy to see that the mapping

$$\mathbb{C}^2 \to \mathbb{C}^2 : (z, w) \mapsto (\overline{z}, w)$$

exchanges the two spreads. Hence they are equivalent. $\qquad\square$

In many cases it is possible to derive (K2) from other conditions. In this respect we have

LEMMA 5.2. *Let $\varphi : \mathbb{C} \to \mathbb{C}$ be a mapping which satisfies (K1). Then $\varphi$ also satisfies (K2) if one of the following conditions is fulfilled.*

(a) *$\varphi$ is a strict contraction, i.e. there exists $c \in [0, 1)$ such that for all $m, n \in \mathbb{C}$ we have $|\varphi(m) - \varphi(n)| \le c|m - n|$.*

(b) *$\varphi$ is bounded.*

*Proof.* (a) We have

$$|m| - |\varphi(m) - \varphi(0)| \ge |m| - c|m - 0| = (1 - c)|m|.$$

This implies

$$|m| - |\varphi(m)| \ge (1 - c)|m| + |\varphi(0)| \to \infty \quad \text{for } |m| \to \infty.$$

(b) This is obvious. $\qquad\square$

In the next section we will show that the translation plane defined by a mapping $\varphi$ as in Theorem 5.1 is desarguesian if and only if $\varphi$ is real linear (Proposition 5.5). Since there exist non-linear mappings $\varphi : \mathbb{C} \to \mathbb{C}$ which satisfy (K1) and (K2), Theorem 5.1 provides a new proof for the fact that non-desarguesian locally compact 4-dimensional translation planes do exist.

If $\varphi : \mathbb{C} \to \mathbb{C}$ satisfies (K1) and (K2), then $\bar{\varphi}$ does as well. So Theorem 5.1 and Lemma 4.4 give a new proof for the fact that locally compact 4-dimensional translation planes can be transposed, cf. section 1.6.

## 5.2  Flocks of Reguli in Topological Spreads

By inspecting Betten's classification of all locally compact 4-dimensional translation planes whose automorphism group is at least 7-dimensional one notices that the majority of the corresponding spreads contains a parabolic flock of reguli. On the other hand, hyperbolic flocks of reguli occur only at one instance, cf. Proposition 5.10. In any case it seems reasonable to investigate topological spreads which contain flocks of reguli more closely.

PROPOSITION 5.3.  *Let $\varphi : \mathbb{C} \to \mathbb{C}$ be a mapping which satisfies the conditions (K1) and (K2) from Theorem 5.1. Furthermore, let $\mathcal{B} = \{S\} \cup \{L(m, \varphi(m))\mid m \in \mathbb{C}\}$ be one of the spreads defined by $\varphi$. Then the following conditions are equivalent:*

*(a) $\mathcal{B}$ contains a parabolic flock of reguli with carrier $S$.*

*(b) The shears group $\Sigma_{[s,S]}$ contains a 1-dimensional subgroup.*

*(c) There are $a, b \in \mathbb{C}, a \neq 0$, such that $\varphi(m + at) = \varphi(m) + bt$ for all $m \in \mathbb{C}, t \in \mathbb{R}$.*

*Proof.* Assume first that $\mathcal{B}$ contains a parabolic flock of reguli with carrier $S$. Then $\Sigma_{[s,S]}$ contains a 1-dimensional subgroup by Proposition 4.17.

Let now $\Delta$ be a 1-dimensional subgroup of $\Sigma_{[s,S]}$. We use the same notation as in section 4.2. All elements of $\Sigma_{[s,S]}$ can be written as

$$\mathbb{C}^2 \to \mathbb{C}^2 : (z, w) \mapsto (z, \lambda_{a,b}(z) + w) \quad \text{where } a, b \in \mathbb{C}.$$

The group of these mappings is isomorphic to the group $\mathbb{R}^4$. Hence, there are $a, b \in \mathbb{C}, (a, b) \neq (0, 0)$, such that $\Delta$ consists of all mappings

$$\chi_t : \mathbb{C}^2 \to \mathbb{C}^2 : (z, w) \mapsto (z, \lambda_{at, bt}(z) + w) \quad \text{with } t \in \mathbb{R}$$

By Lemma 4.3 we have $\chi_t(L(m, \varphi(m))) = L(m + at, \varphi(m) + bt) = L(m + at, \varphi(m + at))$. This implies $a \neq 0$ and we get (c).

Assume now that (c) is satisfied. Then Lemma 4.12 implies that $\mathcal{B}$ contains a parabolic flock of reguli with carrier $S$. $\square$

In concrete examples it is usually possible to take $b = 0$ and $a = 1$ or $a = i$, cf. Proposition 5.7 – 5.9.

PROPOSITION 5.4.  *Let $\varphi : \mathbb{C} \to \mathbb{C}$ be a mapping which satisfies the conditions (K1) and (K2) from Theorem 5.1. Furthermore, let $\mathcal{B} = \{S\} \cup \{L(m, \varphi(m))\mid$*

$m \in \mathbb{C}\}$ *be one of the spreads defined by* $\varphi$. *Assume moreover that* $\varphi(0) = 0$, *i.e.* $W = L(0,0) \in \mathcal{B}$. *Then the following conditions are equivalent:*

(a) $\mathcal{B}$ *contains a hyperbolic flock of reguli with carrier* $(S, W)$.

(b) *We have* $\varphi(ms) = \varphi(m)s$ *for all* $m \in \mathbb{C}, s \in \mathbb{R}$.

*Proof.* Assume that $\mathcal{B}$ contains a hyperbolic flock of reguli with carrier $(S, W)$. Let $L(m, \varphi(m)) \in \mathcal{B}$. By Lemma 4.12, we also have $L(ms, \varphi(m)s) \in \mathcal{B}$ for all $s \in \mathbb{R}$. Hence (b) is satisfied.

The proof of the converse is similar. □

There is no simple characterization of hyperbolic flocks of reguli by the existence of a 1-dimensional group of homologies. In Proposition 5.10 we will study examples of topological spreads which do not contain a hyperbolic flock of reguli with carrier $(S, W)$ although $\Sigma_{[s,W]}$ contains a subgroup isomorphic to $\mathbb{R}^\times$.

We now give a characterization of the complex affine plane.

PROPOSITION 5.5. *Let* $\varphi : \mathbb{C} \to \mathbb{C}$ *be a mapping which satisfies the conditions (K1) and (K2) from Theorem 5.1. Assume moreover that* $\varphi(0) = 0$. *Let* $\mathcal{B} = \{S\} \cup \{L(m, \varphi(m)) \mid m \in \mathbb{C}\}$ *be one of the spreads associated with* $\varphi$ *and let* $\mathcal{A}$ *denote the corresponding translation plane. Then* $\mathcal{A}$ *is desarguesian if and only if* $\varphi$ *is real linear.*

*Proof.* Assume first that $\mathcal{A}$ is desarguesian. This implies in particular that $\Sigma_{[s,S]}$ is a transitive group of shears. Since $\varphi(0) = 0$, Proposition 4.6 implies that $\varphi$ is additive. An additive continuous mapping from $\mathbb{C}$ to $\mathbb{C}$ is necessarily real linear.

Let now $\varphi$ be real linear. Then it follows from 4.8 that $\mathcal{A}$ is pappian and hence also desarguesian. □

In the proof it turned out that there are no locally compact 4-dimensional planes of Lenz type V [88: 1.3'] and that each desarguesian locally compact 4-dimensional affine plane is automatically pappian [87: 7.19].

## 5.3 Locally Compact 4-dimensional Translation Planes with Large Collineation Group

All locally compact 4-dimensional translation planes which admit an at least 7-dimensional collineation group were classified in a series of papers by Betten ([8] – [13]). He describes all these planes using *-transversal homeomorphisms. In this section we give simplified descriptions for some of these planes using $\mathbb{C}_1$-indicator sets. Sometimes, it is also possible to give a simpler treatment of the existence conditions. In particular, we will investigate the planes from [8], [11: Satz 3], [12: Satz 3 and Satz 4] and [13].

Before we go into details we need some preparations.

Betten always considers spreads of the 4-dimensional real vector space $\mathbb{R}^2 \times \mathbb{R}^2$ which contain the element $S = \{0\} \times \mathbb{R}^2$. The other elements of the spread are graphs of linear mappings, which can be described using $2 \times 2$-matrices. These matrices are given by Betten. If we identify $\mathbb{C}$ and $\mathbb{R}^2$ in the usual manner we can transform Betten's description of spreads into ours. Let $m = a + ib, n = c + id \in \mathbb{C}$. The matrix of the linear mapping $\lambda_{m,n} : \mathbb{C} \to \mathbb{C} : z \mapsto mz + n\overline{z}$ with respect to the basis $\{1, i\}$ of $\mathbb{C}$ over $\mathbb{R}$ is given by

$$\begin{pmatrix} a+c & -b+d \\ b+d & a-c \end{pmatrix} = \begin{pmatrix} a_{11} & a_{12} \\ a_{21} & a_{22} \end{pmatrix}.$$

If the matrix on the right is given, one easily computes

(*)
$$a = \frac{1}{2}(a_{11} + a_{22}), \quad b = \frac{1}{2}(a_{21} - a_{12}),$$
$$c = \frac{1}{2}(a_{11} - a_{22}), \quad d = \frac{1}{2}(a_{21} + a_{12}).$$

In Betten's description the matrix coefficients $a_{ij}$ depend on two real parameters $s, t$. Thus, we get $a, b, c$ and $d$ also as functions of $s$ and $t$. Theorem 5.1 says that there exists a mapping $\varphi : \mathbb{C} \to \mathbb{C}$ satisfying (K1) and (K2) such that either $c + id = \varphi(a + ib)$ or $a + ib = \varphi(c + id)$. Hence it is possible to eliminate the parameters $s$ and $t$, at least in principle.

For later use we note the following

LEMMA 5.6. *Let $\varphi : \mathbb{C} \to \mathbb{C}$ be a mapping which satisfies the conditions (K1) and (K2) from Theorem 5.1. Furthermore, let $c_1, c_2 \in \mathbb{C}$ such that $|c_1 c_2| = 1$. We define $\psi : \mathbb{C} \to \mathbb{C}$ by $\psi(m) = c_1 \varphi(c_2 m)$. Then $\psi$ also satisfies (K1) und (K2). The translation planes defined by $\varphi$ and $\psi$ are isomorphic.*

*Proof.* Let $a_1, a_3 \in \mathbb{C}$ with $|a_1| \neq 0 \neq |a_3|$. Consider the real linear mapping

$$\chi : \mathbb{C}^2 \to \mathbb{C}^2 : (z, w) \mapsto (a_1 z, a_3 w).$$

By Lemma 4.3 we have

$$\chi(L(m, \varphi(m)) = L(\frac{a_3}{a_1} m, \frac{a_3}{\overline{a}_1} \varphi(m)).$$

Hence, $\chi(\mathcal{B}_\varphi) = \mathcal{B}_\psi$, where $\psi : \mathbb{C} \to \mathbb{C}$ is given by $\psi(m) = c_1 \varphi(c_2 m), c_1 = \frac{a_3}{\overline{a}_1}, c_2 = \frac{a_1}{a_3}$. From Hilbert's Satz 90 we infer that, given $c_1, c_2 \in \mathbb{C}$ with $|c_1 c_2| = 1$, we can always find $a_1, a_3$ such that $c_1 = \frac{a_3}{\overline{a}_1}$ and $c_2 = \frac{a_1}{a_3}$. $\qquad\square$

PROPOSITION 5.7. *Let $\mu \in \mathbb{R}$ with $0 < |\mu| < 1$. Furthermore, let $\varphi : \mathbb{C} \to \mathbb{C}$ be defined by*

$$\varphi(a + ib) = \begin{cases} 0 & \text{if } b \geq 0 \\ i\mu b & \text{if } b < 0. \end{cases}$$

Then $\mathcal{B}_\mu = \{S\} \cup \{L(m, \varphi(m)) \mid m \in \mathbb{C}\}$ is a topological spread of the 4-dimensional real vector space $\mathbb{C}^2$. The spreads $\mathcal{B}_\mu$ and $\mathcal{B}_{\mu'}$ are equivalent if and only if $\mu' = \pm\mu$. Let $\mathcal{A}_\mu$ be the translation plane associated with $\mathcal{B}_\mu$. The collineation group of $\mathcal{A}_\mu$ is the semidirect product of the translation group $\mathbb{C}^2$ and the group generated by the real linear mappings

$$(z, w) \mapsto (az + bw, cz + dw), a, b, c, d \in \mathbb{R}, ad - bc > 0,$$

$$(z, w) \mapsto (\overline{z}, -\overline{w}) \quad and \quad (z, w) \mapsto (\mu z + \overline{z}, \mu w + \overline{w}).$$

This group is 8-dimensional.

Every non-desarguesian locally compact, 4-dimensional translation plane whose collineation group contains a subgroup isomorphic to $\mathrm{SL}_2(\mathbb{R})$ which acts reducibly on the translation group is isomorphic to one of the planes just described.

The spread $\mathcal{B}_\mu$ contains a parabolic flock of reguli with carrier $S$.

*Proof.* All locally compact, 4-dimensional translation planes whose collineation group contains a subgroup as described in the proposition were determined in [8: Satz 5]. There, the corresponding spreads are given by the matrices

$$\begin{pmatrix} \alpha & \beta \\ -\beta & \alpha \end{pmatrix}, \alpha \in \mathbb{R}, \beta \geq 0 \quad and \quad \begin{pmatrix} \alpha & \beta \\ -w\beta & \alpha \end{pmatrix}, \alpha \in \mathbb{R}, \beta < 0,$$

where $w$ is a real parameter with $w > 0$. The spreads corresponding to $w$ and $w'$ are equivalent if and only if $w' = w$ or $w' = \frac{1}{w}$. Using (*) we compute from the matrices on the left $a = \alpha, b = -\beta$ and $c = d = 0$. From the matrices on the right we get $a = \alpha, b = -\frac{1}{2}(1 + w)\beta, c = 0$ and $d = \frac{1}{2}(1 - w)\beta$. So we have

$$c + id = \varphi(a + ib) = \begin{cases} 0 & if\ b \geq 0 \\ i\mu b & if\ b < 0, \end{cases}$$

where $\mu = \frac{w-1}{w+1}$. A simple computation shows that $\varphi$ is a contraction if and only if $|\mu| < 1$. This condition is obviously equivalent to $w > 0$. If $\varphi$ is a contraction it is automatically a strict contraction. Hence, (K2) is also satisfied by Lemma 5.2. For $w = 1$ or $\mu = 0$, respectively, we get the complex plane.

Since $\varphi$ does not depend on $a$, Proposition 5.3 implies that $\mathcal{B}_\mu$ contains a parabolic flock of reguli with carrier $S$.

The other assertions of the proposition are proved in [8: Satz 5]. $\square$

The planes $\mathcal{A}_\mu$ admit collineations which move $S$. Hence, every spread $\mathcal{B}_\mu$ contains at least one regulus which does not contain $S$. This example provides a negative answer to a question posed by Heimbeck und Wagner [55: p.111].

**PROPOSITION 5.8.** *Let $\gamma \in \mathbb{R}$ with $0 < |\gamma| \leq 1$. Furthermore, let $\varphi : \mathbb{C} \to \mathbb{C}$ be defined by*

$$\varphi(a + ib) = \gamma e^{ia}.$$

Then $B_\gamma = \{S\} \cup \{L(m, \varphi(m)) \mid m \in \mathbb{C}\}$ is a topological spread of the 4-dimensional real vector space $\mathbb{C}^2$. The spreads $B_\gamma$ and $B_{\gamma'}$ are equivalent if and only if $\gamma' = \pm\gamma$. Let $A_\gamma$ be the translation plane associated with $B_\gamma$. The collineation group of $A_\gamma$ is 7-dimensional. It is the semidirect product of the translation group $\mathbb{C}^2$ and the group $\Delta$ generated by the real linear mappings

$$(z, w) \mapsto re^{is}(z, 2(s + it)z + w), r \in \mathbb{R}^\times, s, t \in \mathbb{R} \quad \text{and}$$

$$(z, w) \mapsto (i\bar{z}, -i\bar{w}).$$

The group $\Delta$ fixes $S$ and acts transitively on $B \setminus \{S\}$ as well as on the set of 1-dimensional subspaces of $S$. Moreover, $\Delta$ contains a 1-dimensional group of shears with axis $S$ and center $s$. Conversely, every non-desarguesian locally compact 4-dimensional translation plane whose collineation group contains a subgroup acting the way just described is isomorphic to a translation plane $A_\gamma$.

The spread $B_\gamma$ contains a parabolic flock of reguli with carrier $S$.

*Proof.* All locally compact 4-dimensional translation planes whose collineation group contains a subgroup acting as in the proposition were determined in [13]. There, the corresponding spreads are given by the following matrices

$$\begin{pmatrix} wn\sin s\cos s + n\sin^2 s + cs + wt & n\sin s\cos s - wn\sin^2 s + t \\ wn\cos^2 s + n\sin s\cos s - t & n\cos^2 s - wn\sin s\cos s + cs + wt \end{pmatrix},$$

$s, t \in \mathbb{R}$.

The real parameters $c, w, n$ satisfy the condition $n^2(w^2 + 1)^2 \leq c^2 \neq 0$. Using (∗) one computes

$$a = \frac{1}{2}(2cs + 2wt + n),$$

$$b = \frac{1}{2}(-2t + wn),$$

$$c = \frac{1}{2}(2wn\sin s\cos s + n(\sin^2 s - \cos^2 s))$$

$$= \frac{1}{2}(wn\sin 2s - n\cos 2s),$$

$$d = \frac{1}{2}(2n\sin s\cos s + wn(\cos^2 s - \sin^2 s))$$

$$= \frac{1}{2}(n\sin 2s + wn\cos 2s).$$

The equations for $a$ and $b$ can be solved for $s$ and $t$ giving us

$$t = -b + \frac{1}{2}wn \quad \text{and} \quad s = \frac{1}{c}(a + wb - \frac{1}{2}w^2 n - \frac{1}{2}n).$$

The equations for $c$ and $d$ can be combined to

$$c + id = \frac{n}{2}(-1 + iw)e^{-2is}.$$

Substituting the expression just found for $s$ we get

$$c + id = \varphi(a + ib) = \gamma e^{i\varrho(a+ib)},$$

where

$$\alpha = -\frac{2}{c}, \quad \beta = -2\frac{w}{c}, \quad \gamma = \frac{n}{2}(-1 + iw)e^{in(w^2+1)/c} \quad \text{and} \quad \varrho(a + ib) = \alpha a + \beta b.$$

Betten's condition $n^2(w^2 + 1)^2 \leq c^2$ is equivalent to $|\gamma|^2(\alpha^2 + \beta^2) \leq 1$. We show next that $\varphi$ is a contraction if and only if this condition is satisfied. Since $\varphi$ is bounded, (K2) will be satisfied automatically.

Let $a_1, a_2, b_1, b_2 \in \mathbb{R}$ with $a_1 + ib_1 \neq a_2 + ib_2$. We have to investigate the inequality

$$|\varphi(a_1 + ib_1) - \varphi(a_2 + ib_2)| = |\gamma| \cdot |1 - e^{i\varrho(a_2 - a_1 + i(b_2 - b_1))}| < |a_2 - a_1 + i(b_2 - b_1)|.$$

We put $a = a_2 - a_1$ and $b = b_2 - b_1$ and square both sides of the inequality. This gives us

$$|\gamma|^2 \cdot |1 - e^{i\varrho(a+ib)}|^2 = |\gamma|^2 \cdot |1 - e^{i(\alpha a + \beta b)}|^2 < |a + ib|^2 = a^2 + b^2.$$

We put $t = \alpha a + \beta b, s = -\beta a + \alpha b$. Note that $t^2 + s^2 = (\alpha^2 + \beta^2)(a^2 + b^2)$. Using this relation we get

$$|\gamma|^2(\alpha^2 + \beta^2)|1 - e^{it}|^2 < t^2 + s^2.$$

Since this shall hold for all $s \in \mathbb{R}$, we get the following necessary and sufficient condition for the existence of a translation plane

$$f(t) = 4|\gamma|^2(\alpha^2 + \beta^2)\sin^2\left(\frac{t}{2}\right) - t^2 < 0 \quad \text{for } t \neq 0.$$

We have $f(0) = f'(0) = 0$. Consequently, $f$ must have a local maximum at 0. This gives us the necessary condition

$$f''(0) = 2|\gamma|^2(\alpha^2 + \beta^2) - 2 \leq 0.$$

As $\sin^2\left(\frac{t}{2}\right) < \left(\frac{t}{2}\right)^2$ for $t \neq 0$ this condition is also sufficient.

Let $u \in \mathbb{C}$. Using Lemma 4.3 we see that the mapping

$$(z, w) \mapsto e^{i\varrho(u)}(z, 2uz + w)$$

maps the line $L(m, \gamma e^{i\varrho(m)})$ onto the line $L(m + 2u, \gamma e^{i\varrho(m+2u)})$. Hence, the group of these mappings consists of collineations of the translation plane defined by $\varphi$. In [13: S.139] it is erroneously stated that these groups are non-isomorphic for different values of $c$ and $w$ (or $\alpha$ and $\beta$, respectively). By Lemma 5.6 we get an equivalent spread (and hence also an isomorphic collineation group) if we replace $\varphi$ by $\psi$ with $\psi(m) = c_1 \gamma e^{i\varrho(c_2 m)}$, where $c_1, c_2 \in \mathbb{C}$ with $|c_1 c_2| = 1$. By an appropriate choice of $c_1$ and $c_2$ we can achieve $\varrho(c_2 m) = \text{Re}(m)$ and $c_1 \gamma \in [0, \infty)$. For Betten's parameters this means that, up to isomorphism, we can assume $w = 0, c = 2$ and $n > 0$. After this normalization the spread is given by the following matrices

$$\begin{pmatrix} a + \gamma \cos a & -b + \gamma \sin a \\ b + \gamma \sin a & a - \gamma \cos a \end{pmatrix}, a, b \in \mathbb{R},$$

where $\gamma$ is a real parameter with $0 < |\gamma| \le 1$.

Since $\varphi$ does not depend on $b$, Proposition 5.3 implies that $\mathcal{B}_\gamma$ contains a parabolic flock of reguli with carrier $S$.

The other assertions of the proposition are proved in [13]. $\qquad\square$

PROPOSITION 5.9. *Let* $u, v \in \mathbb{C}, (u, v) \ne (0, 0)$, *with* $|u|^2 \le \frac{1}{5}$ *and* $|v|^2 \le \frac{1}{5}$. *Furthermore, let* $\varphi : \mathbb{C} \to \mathbb{C}$ *be defined by*

$$\varphi(a + ib) = \begin{cases} ube^{2i \log |b|} & \text{if } b < 0 \\ 0 & \text{if } b = 0 \\ vbe^{2i \log |b|} & \text{if } b > 0. \end{cases}$$

*Then* $\mathcal{B}_{u,v} = \{S\} \cup \{L(m, \varphi(m)) \mid m \in \mathbb{C}\}$ *is a topological spread of the 4-dimensional real vector space* $\mathbb{C}^2$. *The spreads* $\mathcal{B}_{u,v}$ *and* $\mathcal{B}_{u',v'}$ *are equivalent if and only if there exists* $c \in \mathbb{C}$ *with* $|c| = 1$ *such that either* $(u', v') = c(u, v)$ *or* $(u', v') = c(v, u)$. *Let* $\mathcal{A}_{u,v}$ *be the translation plane associated with* $\mathcal{B}_{u,v}$. *The collineation group of* $\mathcal{A}_{u,v}$ *is 7-dimensional. For* $u \ne v$ *it is the semidirect product of the translation group* $\mathbb{C}^2$ *and the group* $\Delta$ *of all mappings*

$$(z, w) \mapsto re^{is}(z, tz + e^s w), r \in \mathbb{R}^\times, s, t \in \mathbb{R}.$$

*In case* $u = v$ *there is another collineation given by the mapping* $(z, w) \mapsto (z, -w)$. *The group* $\Delta$ *fixes* $S$ *and contains a 1-dimensional group of shears with axis* $S$ *and center* $s$. *Moreover,* $\Delta$ *acts transitively on the set of 1-dimensional subspaces of* $S$ *but not on* $\mathcal{B} \setminus \{S\}$. *Every non-desarguesian locally compact 4-dimensional translation plane whose collineation group contains a subgroup acting that way is isomorphic to a plane* $\mathcal{A}_{u,v}$.

*The spread* $\mathcal{B}_{u,v}$ *contains a parabolic flock of reguli with carrier* $S$.

*Proof.* All locally compact 4-dimensional translation planes which admit a collineation group as described in the proposition were determined in [11: Satz 3]. There, the corresponding spreads are given by the following matrices

(i) $\qquad\qquad\qquad \begin{pmatrix} t & 0 \\ 0 & t \end{pmatrix}, t \in \mathbb{R},$

(ii)
$$\begin{pmatrix} e^s(-(w+1)\sin s \cos s + z \sin^2 s) + t & e^s(\cos^2 s - z \sin s \cos s - w \sin^2 s) \\ e^s(w \cos^2 s - z \sin s \cos s - \sin^2 s) & e^s(z \cos^2 s + (w+1)\cos s \sin s) + t \end{pmatrix},$$

$s, t \in \mathbb{R},$

(*iii*)

$$\begin{pmatrix} e^s((1-p)\sin s\cos s + q\sin^2 s) + t & e^s(-\cos^2 s - q\sin s\cos s - p\sin^2 s) \\ e^s(p\cos^2 s - q\sin s\cos s + \sin^2 s) & e^s(q\cos^2 s + (p-1)\cos s\sin s) + t \end{pmatrix},$$

$s, t \in \mathbb{R}$.

The real parameters $w, z, p, q$ satisfy the conditions $4(z^2 + (w+1)^2) + z^2 + 4w \leq 0$ and $4(q^2 + (p-1)^2) + q^2 - 4p \leq 0$.

For the matrices of type (i) one computes using (*) that $a = t$ and $b = c = d = 0$. Hence we have

$$\varphi(a + ib) = 0 \quad \text{for } b = 0.$$

For the matrices of type (ii) one gets

$$a = \frac{1}{2}ze^s + t,$$

$$b = \frac{1}{2}e^s \left((w-1)\cos^2 s - (1-w)\sin^2 s\right)$$
$$= \frac{1}{2}(w-1)e^s,$$

$$c = \frac{1}{2}e^s \left(-2(w+1)\sin s\cos s + z(\sin^2 s - \cos^2 s)\right)$$
$$= \frac{1}{2}e^s(-z\cos 2s - (w+1)\sin 2s),$$

$$d = \frac{1}{2}e^s \left((w+1)(\cos^2 s - \sin^2 s) - 2z\sin s\cos s\right)$$
$$= \frac{1}{2}e^s \left((w+1)\cos 2s - z\sin 2s\right).$$

The conditions for $z$ and $w$ imply $b < 0$. The equations for $c$ and $d$ can be combined to

$$c + id = \frac{1}{2}e^s(-z + i(w+1))e^{2is}.$$

The equation for $b$ can be solved for $s$ and by substituting we get

$$c + id = \varphi(a + ib) = ube^{2i\log|b|} \quad \text{for } b < 0,$$

where

$$u = \frac{1}{w-1}(-z + i(w+1))e^{2i\log(2/1-w)}.$$

An analogous computation for the matrices of type (iii) finally gives

$$\varphi(a + ib) = vbe^{2i\log b} \quad \text{for } b > 0$$

where

$$v = \frac{1}{p+1}(-q + i(p-1))e^{2i\log(2/p+1)}.$$

Betten's conditions can be transformed to

$$|u|^2 = \frac{z^2 + (w+1)^2}{(w-1)^2} \le \frac{1}{5} \quad \text{and} \quad |v|^2 = \frac{q^2 + (p-1)^2}{(p+1)^2} \le \frac{1}{5}.$$

It can be shown that $\varphi$ is a contraction if and only if these conditions are satisfied. The argument is similar to the one given in the proof of Proposition 5.8 and will not be presented here.

Let $c_1, c_2 \in \mathbb{C}$ with $|c_1 c_2| = 1$. By Lemma 5.6 the mappings $\varphi$ and $\psi$ with $\psi(m) = c_1\varphi(c_2 m)$ lead to equivalent spreads. Since $\varphi$ does not depend on $a$, only real values are allowed for $c_2$. Let $u'$ and $v'$ be the parameters belonging to $\psi$. Then we have

$$(u', v') = \begin{cases} c_1 c_2 e^{i \log c_2}(u, v) & \text{if } c_2 > 0 \\ c_1 c_2 e^{i \log |c_2|}(v, u) & \text{if } c_2 < 0. \end{cases}$$

So, one can e.g. normalize the parameters to $0 \le u \le \sqrt{1/5}$ and $|v| \le u$.

In [11: S.427] it is claimed that the planes under discussion depend on 4 parameters. For the proof one is referred to the corresponding proof of [11: Satz 2]. There, it was shown that the only mappings one can use for the reduction of parameters are of the form

$$(z, w) \mapsto (z, cw) \quad \text{with } c \in \mathbb{R}^\times.$$

Then it was argued that $c = \pm 1$ are the only possibilities. As we have just seen, this last statement is not true. Furthermore, it turns out that the planes of [11: Satz 2] also depend on one parameter less then claimed there.

Since $\varphi$ does not depend on $a$, it follows from Proposition 5.3 that $\mathcal{B}_{u,v}$ contains a parabolic flock of reguli with carrier $S$.

The other statements of the proposition are proved in [11: Satz 3]. $\square$

PROPOSITION 5.10. *Let $k \in \mathbb{N} \cup \{\digamma\}, \lambda \in \mathbb{R}$ and $u \in \mathbb{C}$ such that $\lambda^2|u|^2 \le (1 - |u|^2)(1 - k^2|u|^2)$ and $0 < |u| < 1$. Furthermore, let $\varphi : \mathbb{C} \to \mathbb{C}$ be defined by*

$$\varphi(m) = u\frac{m^k}{|m|^{k-1}}e^{i\lambda \log |m|}.$$

*Then $\mathcal{B}_{k,\lambda,u} = \{S\} \cup \{L(m, \varphi(m)) \mid m \in \mathbb{C}\}$ is a topological spread of the 4-dimensional real vector space $\mathbb{C}^2$. The spreads $\mathcal{B}_{k,\lambda,u}$ and $\mathcal{B}_{k',\lambda',u'}$ are equivalent if and only if $k' = k, \lambda' = \pm\lambda$ and $|u'| = |u|$. Let $\mathcal{A}_{k,\lambda,u}$ be the translation plane associated with $\mathcal{B}_{k,\lambda,u}$. The collineation group of $\mathcal{A}_{k,\lambda,u}$ is 7-dimensional. For $\lambda \ne 0$ this group is the semidirect product of the translation group $\mathbb{C}^2$ and the group $\Delta$ of all mappings*

$$(z, w) \mapsto re^{i\lambda \log |t|}(t^{k-1}z, t^{k+1}w), r \in \mathbb{R}^\times, t \in \mathbb{C}^\times.$$

In case $\lambda = 0$ there is another collineation given by the mapping $(z, w) \mapsto (u\bar{z}, u\bar{w})$. Let $W = L(0, 0)$. The group $\Delta$ fixes $S$ and $W$ and acts transitively on $B \setminus \{S, W\}$. Moreover, $\Delta$ operates transitively on the set of all 1-dimensional subspaces of $S$ as well as $W$. Every non-desarguesian locally compact 4-dimensional translation plane whose collineation group contains a subgroup acting the way just described is isomorphic to a plane $\mathcal{A}_{k,\lambda,u}$.

The spread $\mathcal{B}_{k,\lambda,u}$ contains a hyperbolic flock of reguli with carrier $(S, W)$ if and only if $\lambda = 0$ and $k$ is odd.

*Proof.* All locally compact 4-dimensional translation planes whose collineation group acts as described in the proposition were determined in [12: Satz 3 and Satz 4]. The planes of Satz 4 correspond to the case $\lambda = 0$.

The spreads in [12: Satz 3] are given by the following matrices

$$e^{qt-ps} \begin{pmatrix} a_{11}(s,t) & a_{12}(s,t) \\ a_{21}(s,t) & a_{22}(s,t) \end{pmatrix}, s, t \in \mathbb{R},$$

where

$$a_{11}(s,t) = \cos s \cos t + c \cos s \sin t + d \sin s \sin t,$$
$$a_{12}(s,t) = -\sin s \cos t - c \sin s \sin t + d \cos s \sin t,$$
$$a_{21}(s,t) = -\cos s \sin t + c \cos s \cos t + d \sin s \cos t,$$
$$a_{22}(s,t) = \sin s \sin t - c \sin s \cos t + d \cos s \cos t.$$

The real parameters $p, q, c, d$ satisfy the following conditions

(1) $\begin{cases} p = q > 0 & \text{and } -1 \leq d < 0 \\ q > 0, p = \frac{k-1}{k+1} q, k = 1, 2, 3, \ldots & \text{and } d > 0 \end{cases}$

(2) $\quad -(q+p)^2 A + (q-p)^2 B - 4AB \geq 0,$

where $A = \dfrac{(d-1)^2 + c^2}{4d}$ and $B = \dfrac{(d+1)^2 + c^2}{4d}$.

In order not to get confused with the parameters $c$ and $d$ we replace the letters $a, b, c$ and $d$ in (∗) by $\alpha, \beta, \gamma$ and $\delta$. Then we get

$$\alpha = \frac{1}{2} e^{qt-ps} \left((1+d)(\cos s \cos t + \sin s \sin t) + c(\cos s \sin t - \sin s \cos t)\right),$$

$$\beta = \frac{1}{2} e^{qt-ps} \left((-1-d)(\sin s \cos t - \cos s \sin t) - c(\sin s \sin t + \cos s \cos t)\right),$$

$$\gamma = \frac{1}{2} e^{qt-ps} \left((1-d)(\cos s \cos t - \sin s \sin t) + c(\cos s \sin t + \sin s \cos t)\right),$$

$$\delta = \frac{1}{2} e^{qt-ps} \left((-1+d)(\sin s \cos t + \cos s \sin t) + c(\cos s \cos t - \sin s \sin t)\right).$$

Using the addition theorems and Euler's formula these equations can be combined to

$$\alpha + i\beta = \frac{1}{2} e^{qt-ps} (1 + d - ic) e^{i(t-s)},$$

$$\gamma + i\delta = \frac{1}{2} e^{qt-ps} (1 - d + ic) e^{-i(t+s)}.$$

According to Theorem 5.1 there exists a mapping $\varphi : \mathbb{C} \to \mathbb{C}$ satisfying (K1) and (K2) such that either $\alpha(t,s) + i\beta(t,s) = \varphi(\gamma(t,s) + i\delta(t,s))$ or $\gamma(t,s) + i\delta(t,s) = \varphi(\alpha(t,s) + i\beta(t,s))$.

Assume first that $p = q > 0$. We put $r = t - s$. Then we get

$$\alpha + i\beta = \frac{1}{2}(1 + d - ic)e^{pr} e^{ir},$$

$$\gamma + i\delta = \frac{1}{2}(1 - d + ic)e^{pr} e^{-i(2s+r)}.$$

From the second equation we deduce

$$r = \frac{1}{2p}\left(\log 4(\gamma^2 + \delta^2) - \log((1-d)^2 + c^2)\right).$$

Substituting this into the first equation finally gives us

$$\alpha + i\beta = \varphi(\gamma + i\delta) = u|\gamma + i\delta|e^{i\lambda|\gamma + i\delta|},$$

where

$$\lambda = \frac{1}{p} \quad \text{and} \quad u = \frac{1 + d - ic}{\sqrt{(1-d)^2 + c^2}}e^{i\frac{1}{2p}(\log 4 - \log((1-d)^2 + c^2))}.$$

Hence, $\varphi$ is of the form given in the proposition with $k = 0$ and $\lambda \neq 0$. The condition $\lambda^2|u|^2 \leq (1 - |u|^2)(1 - k^2|u|^2)$ is equivalent to

$$\frac{1}{p^2} \cdot \frac{(1+d)^2 + c^2}{(1-d)^2 + c^2} \leq \left(1 - \frac{(1+d)^2 + c^2}{(1-d)^2 + c^2}\right) = \frac{-4d}{(1-d)^2 + c^2}.$$

Since $d < 0$, this condition is equivalent to

$$\frac{(1+d)^2 + c^2}{4d} \geq -p^2.$$

Again from $d < 0$ we deduce that this is equivalent to Betten's condition (2)

$$-(2p)^2 \cdot \frac{(1-d)^2 + c^2}{4d} - 4\frac{(1-d)^2 + c^2}{4d} \cdot \frac{(1+d)^2 + c^2}{4d} \geq 0.$$

From $d < 0$ we also infer $|u| < 1$. We have $\mathcal{B} \setminus \{S\} = \{L(\varphi(m), m) \,|\, m \in \mathbb{C}\}$. From Theorem 5.1 we know that $\{S\} \cup \{L(m, \varphi(m)) \,|\, m \in \mathbb{C}\}$ is also a spread which is equivalent to $\mathcal{B}$.

Assume now that $p = \frac{k-1}{k+1}q, k = 1, 2, 3, \ldots$. Then we have $q - p = \frac{2}{k+1}q$. We put $r = t - s$. This gives us

$$\alpha + i\beta = \frac{1}{2}e^{(q-p)s + qr}(1 + d - ic)e^{ir},$$

$$\gamma + i\delta = \frac{1}{2}e^{(q-p)s + qr}(1 + d - ic)e^{-i(2s+r)}.$$

We put $v = (q - p)s + qr$. Since $q - p = \frac{2}{k+1}q$, this yields

$$\alpha + i\beta = \frac{1}{2}e^v(1 + d - ic)e^{ir},$$

$$\gamma + i\delta = \frac{1}{2}e^v(1 + d - ic)e^{-ir}e^{i(k+1)(r + \frac{1}{q}v)}.$$

The first of these equations implies

$$v = \log 2|\alpha + i\beta| - \log \sqrt{(1 + d)^2 + c^2} \quad \text{and}$$

$$e^{ikr} = 2^k(\alpha + i\beta)^k e^{-kv}(1 - d + ic)^{-k}.$$

Substituting this into the second equation we finally get

$$\gamma + i\delta = \varphi(\alpha + i\beta) = u\frac{(\alpha + i\beta)^k}{|\alpha + i\beta|^{k-1}}e^{i\lambda \log |\alpha + i\beta|},$$

where $\lambda = \frac{k+1}{q}$ and

$$u = \frac{(1 - d + ic)\sqrt{(1 + d)^2 + c^2}^{k-1}}{(1 + d - ic)^k}e^{i\frac{k+1}{q}(\log 2 - \log\sqrt{(1+d)^2+c^2})}.$$

The condition $\lambda^2|u|^2 \leq (1 - |u|^2)(1 - k^2|u|^2)$ is equivalent to

$$\frac{(k + 1)^2}{q^2} \cdot \frac{(1 - d)^2 + c^2}{(1 + d)^2 + c^2} \leq \left(1 - \frac{(1 - d)^2 + c^2}{(1 + d)^2 + c^2}\right)\left(1 - k^2 \cdot \frac{(1 - d)^2 + c^2}{(1 + d)^2 + c^2}\right).$$

Multiplying by $q^2((1 + d)^2 + c^2)^2$ we get

$$(k + 1)^2((1 - d)^2 + c^2)((1 + d)^2 + c^2) \leq 4dq^2\left((1 + d)^2 + c^2 - k^2((1 - d)^2 + c^2)\right).$$

In consideration of $p + q = \frac{2k}{k+1}q$ and $q - p = \frac{2}{k+1}q$, Betten's condition (2) can be written as

$$-\frac{4k^2q^2}{(k + 1)^2} \cdot \frac{(d - 1)^2 + c^2}{4d} + \frac{4q^2}{(k + 1)^2} \cdot \frac{(d + 1)^2 + c^2}{4d}$$

$$- 4\frac{(d - 1)^2 + c^2}{4d} \cdot \frac{(d + 1)^2 + c^2}{4d} \geq 0.$$

Obviously, the last two conditions are equivalent. Since $d > 0$, we also have $|u| < 1$.

The spreads in [12: Satz 4] are given by the following matrices

$$\begin{pmatrix} a_{11}(s, t) & a_{12}(s, t) \\ a_{21}(s, t) & a_{22}(s, t) \end{pmatrix}, s, t \in \mathbb{R},$$

where

$$a_{11}(s,t) = s(\cos nt \cos mt + c\sin nt \cos mt + d\sin nt \sin mt),$$
$$a_{12}(s,t) = s(-\cos nt \sin mt - c\sin nt \sin mt + d\sin nt \cos mt),$$
$$a_{21}(s,t) = s(-\sin nt \cos mt + c\cos nt \cos mt + d\cos nt \sin mt),$$
$$a_{22}(s,t) = s(\sin nt \sin mt - c\cos nt \sin mt + d\cos nt \cos mt).$$

The parameters $n, m \in \mathbb{Z}$, $(m,n) = 1$, and $c, d \in \mathbb{R}$ satisfy the following conditions

$$(1) \quad \begin{cases} m = n = 1 & \text{and } -1 \leq d < 0 \\ m = 1,2,3,\ldots & n = m+1 \quad \text{and } d > 0 \\ m = 1,3,5,\ldots & n = m+2 \quad \text{and } d > 0 \end{cases}$$

$$(2) \quad (n-m)^2 B \geq (n+m)^2 A,$$

$$\text{where } A = \frac{(d-1)^2 + c^2}{4d} \text{ and } B = \frac{(d+1)^2 + c^2}{4d}.$$

Using $(*)$ we compute

$$\alpha = \frac{1}{2}s((1+d)(\cos nt \cos mt + \sin nt \sin mt)$$
$$+ c(\sin nt \cos mt - \cos nt \sin mt)),$$

$$\beta = \frac{1}{2}s(-(1+d)(\cos nt \sin mt - \sin nt \cos mt)$$
$$- c(\sin nt \sin mt + \cos nt \cos mt)),$$

$$\gamma = \frac{1}{2}s((1-d)(\cos nt \cos mt - \sin nt \sin mt)$$
$$+ c(\sin nt \cos mt + \cos nt \sin mt)),$$

$$\delta = \frac{1}{2}s((-1+d)(\cos nt \sin mt + \sin nt \cos mt)$$
$$- c(\sin nt \sin mt - \cos nt \cos mt)).$$

These equations can be combined to

$$\alpha + i\beta = \frac{1}{2}s(1 + d - ic)e^{i(n-m)t},$$
$$\gamma + i\delta = \frac{1}{2}s(1 - d + ic)e^{-i(n+m)t}.$$

A calculation similar to the one just given shows that these spreads lead to the functions $\varphi$ of the proposition with $\lambda = 0$.

It can be shown that the mapping $\varphi$ satisfies (K1) and (K2) if and only if the parameters $k, \lambda$ and $u$ satisy the conditions given in the proposition. However, there seems to be no essential simplification of the calculation given in [12: Satz 3 and Satz 4].

Let $r \in \mathbb{R}^\times$ and $t \in \mathbb{C}^\times$. From Lemma 4.3 we infer that the mapping

$$(z, w) \mapsto re^{i\lambda \log |t|}(t^{k-1}z, t^{k+1}w)$$

maps the line $L(m, \varphi(m))$ to $L(m', n')$ where

$$(m', n') = \frac{re^{i\lambda \log |t|}t^{k+1}}{r^2 |t|^{2(k-1)}}(m, \varphi(m)) \begin{pmatrix} re^{-i\lambda \log |t|}\bar{t}^{k-1} & 0 \\ 0 & re^{i\lambda \log |t|}t^{k-1} \end{pmatrix}$$

$$= (t^2 m, \varphi(t^2 m)).$$

Consequently, this mapping induces a collineation of $\mathcal{A}_{k,\lambda,u}$. Let $\Delta$ denote the group of these mappings.

In [12: p. 142] it is claimed that the group $\Delta$ is its own normalizer in $GL_4(\mathbb{R})$. However, it is easy to see that the normalizer of $\Delta$ consists of all mappings $(z, w) \mapsto (a_1 z, a_3 w)$ with $a_1, a_3 \in \mathbb{C}^{\times}$. As discussed in the proof of Lemma 5.6, the application of such a mapping replaces $\varphi$ by $\psi$ where $\psi(m) = c_1 \varphi(c_2 m), c_1, c_2 \in \mathbb{C}, |c_1 c_2| = 1$. We have

$$c_1 \varphi(c_2 m) = u \frac{c_1 c_2^k}{|c_2|^{k-1}} e^{i\lambda \log |c_1|} \frac{m^k}{|m|^{k-1}} e^{i\lambda \log |m|}.$$

Hence, we get an equivalent spread if we replace $u$ by $cu$ where $|c| = 1$.

Using Lemma 4.3 it is easily seen that the mapping

$$(z, w) \mapsto (u\bar{z}, u\bar{w})$$

maps the line $L(m, u\frac{m^k}{|m|^{k-1}}e^{i\lambda \log |m|})$ to $L(\overline{m}, u\frac{\overline{m}^k}{|m|^{k-1}}e^{-i\lambda \log |m|})$. Hence, the planes $\mathcal{A}_{k,\lambda,u}$ and $\mathcal{A}_{k,-\lambda,u}$ are isomorphic. Moreover, the mapping just given induces a collineation of $\mathcal{A}_{k,0,u}$. In [12] it is shown that these are in fact all collineations.

By Proposition 5.4, the spread $\mathcal{B}_{k,\lambda,u}$ contains a hyperbolic flock of reguli with carrier $(S, W)$ if and only if $\varphi(ms) = \varphi(m)s$ for all $m \in \mathbb{C}$ and $s \in \mathbb{R}$. Obviously, this condition is satisfied if and only if $\lambda = 0$ and $k$ is odd. $\qquad\square$

# 6. Planes of Lenz Type V with Complex Kernel

In this chapter we determine all locally compact planes of Lenz type V whose kernel contains a subfield isomorphic to $\mathbb{C}$. Apart from the complex plane, such a plane is necessarily of topological dimension 8.

## 6.1 Locally Compact Translation Planes with Complex Kernel

In this section we investigate locally compact connected translation planes whose kernel contains a subfield isomorphic to $\mathbb{C}$. According to Proposition 1.29 planes of this type are associated with topological spreads of a complex vector space $V$ of (complex) dimension 2 or 4. In case $\dim V = 2$ we get the complex affine plane. So from now on we assume $\dim V = 4$. As topological spaces these planes are then 8-dimensional. Since $\mathbb{C}$ is algebraically closed and hence does not admit a quadratic extension field, we will describe these planes using the ring of double numbers over $\mathbb{C}$, cf. section 2.7. In the sequel this ring will be denoted by $A = A(\mathbb{C})$. According to Proposition 2.7 every locally compact 8-dimensional translation plane whose kernel contains a subfield isomorphic to $\mathbb{C}$ can be obtained from an $A_1$-indicator set. First we have to determine under which circumstances the translation plane associated with an $A_1$-indicator set is a topological plane.

We use the notation from section 4.2 with $F = \mathbb{C}$ and $E = A = A(\mathbb{C})$. The spread associated with an $A_1$-indicator set $\mathcal{J} = \{(\varphi(m), \psi(m)) \mid m \in A\}$ is denoted by $\mathcal{B}_{\varphi,\psi}$ and the corresponding translation plane by $\mathcal{A}_{\varphi,\psi}$.

PROPOSITION 6.1.    Let $\mathcal{J} = \{(\varphi(m), \psi(m)) \mid m \in A\}$ be an $A_1$-indicator set and let $\mathcal{A}_{\varphi,\psi}$ denote the translation plane associated with $\mathcal{J}$. Then the following conditions are equivalent:

(a) The mappings $\varphi \circ (\varphi + \psi)^{-1}$ and $\psi \circ (\varphi + \psi)^{-1}$ are continuous.

(a) There exists a mapping $\varrho : \mathbb{C} \to \mathbb{C}$ such that $\varphi \circ \varrho$ and $\psi \circ \varrho$ are continuous.

(c) $\mathcal{A}_{\varphi,\psi}$ is a topological translation plane.

*Proof.* By Lemma 1.15 we may assume that $\mathcal{J}$ is normalized, i.e. $L(0,0) = \{(z,0) \in A^2 \mid z \in A\}$ and $L(1,0) = \{(z,z) \in A^2 \mid z \in A\}$ are contained in $\mathcal{B}_{\varphi,\psi}$.

Obviously, (a) implies (b).

Assume now that there exists $\varrho : A \to A$ such that $\varphi \circ \varrho$ and $\psi \circ \varrho$ are continuous. As in the proof of Proposition 4.6 we may assume that $\varphi$ and $\psi$ themselves are continuous. Since (H2) is satisfied, $\varphi + \psi$ is bijective and hence a homeomorphism by Brouwer's theorem.

Put $\varphi' = \varphi \circ (\varphi + \psi)^{-1}$ and $\psi' = \psi \circ (\varphi + \psi)^{-1}$. Then we have $\mathcal{J} = \{(\varphi'(m), \psi'(m)) \mid m \in A\}$. We define a multiplication $* : A \times A \to A$ by

$$m * z = \varphi'(m)z + \psi'(m)\overline{z}.$$

By Proposition 4.5, the structure $(A, +, *)$ is a left quasifield which coordinatizes $\mathcal{A}_{\varphi,\psi}$. Since $\varphi'$ and $\psi'$ are continuous, the multiplication $*$ is continuous as well. It follows from [45: Satz 2.3] that $(A, +, *)$ is a topological quasifield and by [87: Theorem 7.15] $\mathcal{A}_{\varphi,\psi}$ is a topological affine plane.

Let now $\mathcal{A}_{\varphi,\psi}$ be a topological affine plane. We put $\varrho = (\varphi + \psi)^{-1}$. By the construction given above, $(A, +, *)$ is a topological quasifield. This implies continuity of $\varphi \circ \varrho$ and $\psi \circ \varrho$. $\qquad\square$

## 6.2 Locally Compact Planes of Lenz Type V with Complex Kernel

In this section we give a description of the locally compact projective planes of Lenz type V whose kernel is isomorphic to $\mathbb{C}$.

According to Proposition 1.29 such a plane is either isomorphic to the complex affine plane or it can be constructed from an $A_1$-indicator set as in the last section.

Since the polynomial $P(x) = x^2 + 1 \in \mathbb{C}[x]$ has two different roots in $\mathbb{C}$, the ring $A$ is isomorphic to $\mathbb{C}[x]/(x^2 + 1)$. Hence there exists a $\mathbb{C}$-basis $\{1, j\}$ of $A$ with $j^2 = -1$. We will use this basis from now on and we will identify the element $a + jb \in A$ with $(a, b)^t \in \mathbb{C}^2$ if this is appropriate.

The automorphism $\overline{\phantom{x}} : A \to A$ exchanges $j$ and $-j$, and hence we have $\overline{a + jb} = a - jb$ for all $a, b \in \mathbb{C}$.

**LEMMA 6.2.** *Let $\mathcal{J} = \{(\varphi(m), \psi(m)) \mid m \in A\}$ be an $A_1$-indicator set. Furthermore, let $\varphi$ and $\psi$ be continuous and assume that $\varphi(0) = \psi(0) = 0$. Let $\mathcal{A}_{\varphi,\psi}$ be the translation plane associated with $\mathcal{J}$. Then the following conditions are equivalent:*

(a) *The mappings $\varphi \circ (\varphi + \psi)^{-1}$ and $\psi \circ (\varphi + \psi)^{-1}$ are real linear.*

(b) *There exists a homeomorphism $\varrho : A \to A$ such that $\varphi \circ \varrho$ and $\psi \circ \varrho$ are real linear.*

(c) *The shear group $\Sigma_{[s,S]}$ of $\mathcal{A}_{\varphi,\psi}$ is linearly transitive.*

*Proof.* Let (a) be satisfied. Since $\varphi$ and $\psi$ are continuous, $\varphi+\psi$ is continuous as well. By Brouwer's theorem, $\varphi+\psi$ then is a homeomorphism. So, (b) is satisfied with $\varrho = (\varphi+\psi)^{-1}$.

Assume now that $\varphi \circ \varrho$ and $\psi \circ \varrho$ are real linear, where $\varrho : A \rightarrow A$ is a homeomorphism. Then Proposition 4.6 implies that $\Sigma_{[s,s]}$ is a transitive group of shears of $A_{\varphi,\psi}$.

Let now $\Sigma_{[s,S]}$ be a transitive group of shears. Put $\varphi' = \varphi \circ (\varphi+\psi)^{-1}$ and $\psi' = \psi \circ (\varphi+\psi)^{-1}$.

By Proposition 4.6, $\varphi'$ and $\psi'$ are additive homomorphisms of $A$. Since $\varphi$ and $\psi$ are continuous, $\varphi'$ and $\psi'$ are continuous as well. Every continuous additive homomorphism of a finite dimensional real vector space is real linear.□

Because of Lemma 6.2 we will now investigate real linear mappings of the 2-dimensional complex vector space $A$.

Since we use the bar already for the canonical $\mathbb{C}$-automorphism of $A$, we will denote the usual conjugation of $\mathbb{C}$ with a star. Moreover, we extend the complex conjugation to a ring automorphism of $A$ by defining $(a+jb)^* = a^* + jb^*$ for $a, b \in \mathbb{C}$. It is easy to see that this really defines a ring automorphism of $A$. Also, the ring automorphisms $^*$ and $^-$ commute.

LEMMA 6.3. *Let* $\Lambda : \mathbb{C}^2 \rightarrow \mathbb{C}^2$ *be real linear. Then there are uniquely determined matrices* $B, C \in M_{2,2}(\mathbb{C})$ *such that* $\Lambda = \Lambda_{B,C}$, *where* $\Lambda_{B,C} : \mathbb{C}^2 \rightarrow \mathbb{C}^2$ *is given by*

$$\Lambda_{B,C}(z) = Bz + Cz^*.$$

*If* $\Lambda_{B,C}$ *and* $C$ *are invertible we have* $\Lambda_{B,C}^{-1} = \Lambda_{\hat{B},\hat{C}}$ *where* $\hat{C} = (C^* - B^* C^{-1} B)^{-1}$ *and* $\hat{B} = -C^{*-1}B^*\hat{C}^*$. *If* $B$ *is also invertible, then we have* $\hat{B} = (B - CB^{*-1}C^*)^{-1}$.

*Proof.* The set of all real linear endomorphisms of $\mathbb{C}^2$ and the set of all pairs of complex $2 \times 2$-matrices are both 16-dimensional real vector spaces and $\Lambda : (B,C) \mapsto \Lambda_{B,C}$ is a real linear mapping between these two vector spaces. So it is sufficient to show that $\Lambda_{B,C}$ is the zero mapping if and only if $B = C = 0$. Similar as in the proof of Proposition 2.7, this can be seen by setting $z = (1,0), (i,0), (0,1)$ and $(0,i)$.

Assume now that $C$ and $\Lambda_{B,C}$ are invertible and let $\Lambda_{B,C}^{-1} = \Lambda_{\hat{B},\hat{C}}$. We get $\Lambda_{\hat{B},\hat{C}}$ by solving the system of equations

$$B\hat{B} + C\hat{C}^* = I \quad \text{and} \quad B\hat{C} + C\hat{B}^* = 0.$$

Since $C$ is invertible, the second equation yields

$$\hat{B}^* = -C^{-1}B\hat{C}.$$

Substituting this expression into the first equation leads to

$$-BC^{*-1}B^*\hat{C}^* + C\hat{C}^* = I.$$

Thus we get $\hat{B}$ and $\hat{C}$ as asserted in the lemma.

Assume now that $B$ is also invertible. Then we get

$$
\begin{aligned}
\hat{B} &= -C^{*-1}B^*\hat{C}^* \\
&= -C^{*-1}(B^{*-1})^{-1}(C - BC^{*-1}B^*)^{-1} \\
&= -(CB^{*-1}C^* - BC^{*-1}B^*B^{*-1}C^*)^{-1} \\
&= -(CB^{*-1}C^* - B)^{-1}.
\end{aligned}
$$

$\square$

Let now $\mathcal{B} = \{S\} \cup \{L(\varphi(m), \psi(m)) \mid m \in A\}$ be a topological spread of $A^2$ such that the group of shears $\Sigma_{[s,S]}$ of $\mathcal{A}_{\varphi,\psi}$ is transitive. By Lemma 6.2 we can assume that $\varphi$ and $\psi$ are real linear. We replace $\varphi$ and $\psi$ by $\varphi \circ (\varphi + \psi)^{-1}$ and $\psi \circ (\varphi + \psi)^{-1}$. Then we have $\varphi(m) + \psi(m) = m$ for all $m \in A$. By Lemma 6.3 there are matrices $B, C \in M_{2,2}(\mathbb{C})$ such that $\varphi(m) = \frac{1}{2}(m + Bm + Cm^*)$ and $\psi(m) = \frac{1}{2}(m - Bm - Cm^*)$ for all $m \in A$. Now the question arises under which conditions on $B$ and $C$ the mappings $\varphi$ and $\psi$ define an $A_1$-indicator set. The answer is given by the following

LEMMA 6.4. *Let $B$ and $C$ be complex $2 \times 2$-matrices. Let $\varphi, \psi : A \to A$ be defined by $\varphi(m) = \frac{1}{2}(m + Bm + Cm^*)$ and $\psi(m) = \frac{1}{2}(m - Bm - Cm^*)$. Then $\mathcal{J} = \{(\varphi(m), \psi(m)) \mid m \in A\}$ is an $A_1$-indicator set if and only if*

$$
|m^t Bm| < |m^t Cm^*| \quad \text{for all } m \in A \setminus \{0\}.
$$

*Proof.* By Proposition 2.7 $\mathcal{J}$ is an $A_1$-indicator set if and only if $\varphi$ and $\psi$ satisfy the conditions (H1) and (H2). Since $\varphi$ and $\psi$ are additive, it is sufficient to check (H1) for $n = 0$. If (H1) is satisfied then all mappings $\varrho_z, z \in A \setminus \{0\}$ that appear in (H2) are injective. Since $A$ is a finite dimensional vector space, these mappings are actually bijective. Hence, $\mathcal{J}$ is an $A_1$-indicator set if and only if we have $N(\varphi(m)) \neq N(\psi(m))$ for all $m \in A \setminus \{0\}$.

In the chosen basis of $A$ we have $N(a + jb) = (a + jb)(a - jb) = a^2 + b^2$. Since we identify $A$ and $\mathbb{C}^2$, we can also write $N(m) = m^t m$. Thus we get

$$
N(\varphi(m)) = \frac{1}{4} \left( N(m) + m^t(Bm + Cm^*) + N(Bm + Cm^*) \right),
$$

$$
N(\psi(m)) = \frac{1}{4} \left( N(m) - m^t(Bm + Cm^*) + N(Bm + Cm^*) \right).
$$

It follows that $\mathcal{J}$ is an $A_1$-indicator set if and only if there holds

$$
m^t Bm \neq -m^t Cm^* \quad \text{for all } m \in \mathbb{C}^2 \setminus \{0\}.
$$

Let now $\mathcal{J}$ be an $A_1$-indicator set. We claim that

$$
|m^t Bm| \neq |m^t Cm^*| \quad \text{for } m \neq 0.
$$

Assume that there exists $m \in \mathbb{C}^2 \setminus \{0\}$ with $|m^t Bm| = |m^t Cm^*|$. Then we can find $s \in \mathbb{R}$ such that $e^{2is} m^t Bm = -m^t Cm^*$. This yields $(e^{is} m)^t B(e^{is} m) = -(e^{is} m)^t C(e^{is} m)^*$, and we get a contradiction.

Since $\mathbb{C}^2 \setminus \{0\}$ is connected, it follows that we have either

$$(I) \qquad\qquad |m^t Bm| < |m^t Cm^*| \quad \text{for all } m \in \mathbb{C}^2 \setminus \{0\}$$

or

$$(II) \qquad\qquad |m^t Bm| > |m^t Cm^*| \quad \text{for all } m \in \mathbb{C}^2 \setminus \{0\}.$$

Since every quadratic form on $\mathbb{C}^2$ admits non-trivial isotropic vectors, case (II) is not possible. So we get the condition stated in the lemma.

It is clear that this condition is also sufficient. $\qquad\qquad\qquad\qquad\square$

Now we can formulate the principal result of this section.

PROPOSITION 6.5. *Let $B, C$ be complex $2 \times 2$-matrices such that*

$$(*) \qquad\qquad |m^t Bm| < |m^t Cm^*| \quad \text{for all } m \in A \setminus \{0\}.$$

*Define $\varphi, \psi : A \to A$ by*

$$\varphi(m) = \frac{1}{2}(m + Bm + Cm^*) \quad \text{and} \quad \psi(m) = \frac{1}{2}(m - Bm - Cm^*).$$

*Then $\mathcal{J} = \{(\varphi(m), \psi(m)) \mid m \in A\}$ is an $A_1$-indicator set. The translation plane $A_{B,C}$ associated with $\mathcal{J}$ is a locally compact 8-dimensional plane of Lenz type at least V and the kernel of $A_{B,C}$ contains a subfield isomorphic to $\mathbb{C}$. Conversely, every locally compact 8-dimensional plane of Lenz type at least V whose kernel contains a subfield isomorphic to $\mathbb{C}$ is isomorphic to a plane $A_{B,C}$.*

*Proof.* Let $\varphi, \psi : A \to A$ be given as in the proposition. It follows from Lemma 6.2 – 6.4 that $A_{B,C}$ is a locally compact 8-dimensional plane of Lenz type at least V. By construction, the kernel of $A_{B,C}$ contains a subfield isomorphic to $\mathbb{C}$.

Let now $A$ be a locally compact 8-dimensional plane of Lenz type at least V whose kernel contains a subfield isomorphic to $\mathbb{C}$. Then we may assume that $A$ is associated with a topological spread $\mathcal{B}$ of the 4-dimensional complex vector space $A^2$. By Lemma 1.15 we may further assume that $\{S, L(0,0)\} \subset \mathcal{B}$ and that $\Sigma_{[s,s]}$ is a transitive group of shears. Then we have $\mathcal{B} = \{S\} \cup \{L(\varphi(m), \psi(m)) \mid m \in A\}$, where $\mathcal{J} = \{(\varphi(m), \psi(m)) \mid m \in A\}$ is an $A_1$-indicator set. By Lemma 6.2 we can assume that $\varphi$ und $\psi$ are real linear. After replacing $\varphi$ and $\psi$ by $\varphi \circ (\varphi + \psi)^{-1}$ and $\psi \circ (\varphi + \psi)^{-1}$ we can also assumme that $\varphi(m) + \psi(m) = m$ for all $m \in A$. From Lemma 6.3 we infer that $\varphi$ and $\psi$ can be written as in the proposition and by Lemma 6.4 the proof is complete. $\qquad\qquad\square$

If $A_{B,C}$ is not of Lenz type V, then $A_{B,C}$ is necessarily desarguesian and hence isomorphic to the plane over the skewfield of quaternions. A more careful analysis shows that this happens precisely if $B$ is skew symmetric and $kC$ is hermitian for a suitable $k \in \mathbb{C}$. There is no obvious characterization of the mappings $\varphi$ and $\psi$ arising in this particular case.

Our next aim is to classify the locally compact 8-dimensional planes of Lenz type V whose kernel contains a subfield isomorphic to $\mathbb{C}$.

THEOREM 6.6. *Let $\mathcal{A}$ be a locally compact 8-dimensional plane of Lenz type at least V whose kernel contains a subfield isomorphic to $\mathbb{C}$. Then $\mathcal{A}$ is isomorphic to a plane $\mathcal{A}_{B,C}$ where $B$ and $C$ belong to one of the following families:*

$(I_\delta)$
$$B = 0, \quad C = \begin{pmatrix} 1 & \\ & e^{i\delta} \end{pmatrix}, \ 0 \le \delta < \pi$$

*or*

$(II_{r,c})$
$$B = \begin{pmatrix} 0 & \\ & 1 \end{pmatrix}, \quad C = \begin{pmatrix} 1 & r \\ & c \end{pmatrix}, \ r \in \mathbb{R}, c \in \mathbb{C}, r \ge 0, \operatorname{Im} c \ge 0$$
*and $0 < P_{r,c}(x) = x^4 + (2\operatorname{Re} c - r^2)x^2 - 2rx + |c|^2 - 1$ for all $x \in \mathbb{R}$.*

The proof of this theorem is given in the next section.

Necessary and sufficient conditions for the existence of real roots of the polynomial $P_{r,c}$ can be found in Weber [101: §78].

Our choice of $B$ and $C$ has the useful consequence that the corresponding $A_1$-indicator set $\mathcal{J}$ is normalized. By Proposition 4.5, this implies that $(A, +, \circ)$, where $m \circ z = \varphi(m)z + \psi(m)\bar{z}$ for $m, z \in A$, is a semifield which coordinatizes $\mathcal{A}_{B,C}$. After some calculation one finds the following explicit expressions for the product $m \circ z, m = (m_1, m_2), z = (z_1, z_2) \in A = \mathbb{C}^2$, in the different cases:

$(I_\delta)$
$$(m_1, m_2) \circ (z_1, z_2) = (m_1 z_1 - e^{i\delta} m_2^* z_2, m_1^* z_2 + m_2 z_1)$$

$(II_{r,c})$ $\quad (m_1, m_2) \circ (z_1, z_2) = (m_1 z_1 - c m_2^* z_2 - m_2 z_2, m_1^* z_2 + m_2 z_1 + r m_2^* z_2)$

The quaternion plane arises in case $(I_0)$. In the other cases the dimensions of the respective collineation groups $\Gamma$ are as follows:

$(I_\delta)$
$$\dim \Gamma = 17 \text{ for } \delta \ne 0$$

$(II_{r,c})$
$$\dim \Gamma = \begin{cases} 18 & \text{if } r = 0 \text{ and } \operatorname{Im} c = 1 \\ 16 & \text{if } r = 0 \text{ and } \operatorname{Im} c \ne 1 \\ 15 & \text{if } r \ne 0 \end{cases}$$

Theorem 6.6 extends results of Rees [86], who determined the 4-dimensional real division algebras having the property that at least two of their nuclei are isomorphic to $\mathbb{C}$. These division algebras correspond to the case $(I_\delta), \delta \ne 0$.

The locally compact 8-dimensional translation planes with an at least 17-dimensional collineation group were classified in a series of papers by Hähl [47]. It follows from this classification that the locally compact 8-dimensional translation planes with an 18-dimensional collineation group are exactly the planes belonging to case $(II_{0,c})$, $\operatorname{Im} c = 1$, as well as their dual planes and the transposed planes of these duals [47: Klassifikationssatz (ii)]. As already mentioned, the case $(I_\delta)$, $\delta \neq 0$, leads to the planes over the division algebras of Rees [86; 47: Klassifikationssatz (iii)(4)]. The other planes constructed in this section were not known up to now.

## 6.3 Proof of Theorem 6.6

The proof of Theorem 6.6 is given in a series of lemmas.

**LEMMA 6.7.** *Let $a, b \in A$ and let $D$ be the matrix of $\lambda_{a,b}$.*

*(a) If $D = \begin{pmatrix} p & r \\ s & q \end{pmatrix}$ then $a = \frac{1}{2}(p+q+j(s-r))$ and $b = \frac{1}{2}(p-q+j(s+r))$.*

*(b) The matrix of $\lambda_{\bar{a},b}$ is $D^t$.*

*Proof.* Let $a = a_1 + ja_2$ and $b = b_1 + jb_2$, $a_i, b_i \in \mathbb{C}$. Then we get $\lambda_{a,b}(1) = a + b = a_1 + b_1 + j(a_2 + b_2)$ and $\lambda_{a,b}(j) = aj - bj = -a_2 + b_2 + j(a_1 - b_1)$. It follows that

$$D = \begin{pmatrix} p & r \\ s & q \end{pmatrix} = \begin{pmatrix} a_1 + b_1 & -a_2 + b_2 \\ a_2 + b_2 & a_1 - b_1 \end{pmatrix}.$$

Solving this equation proves (a). Since exchanging $r$ and $s$ replaces $a$ by $\bar{a}$ and leaves $b$ invariant, we also get (b). □

Let $J \in \mathrm{GL}_2(\mathbb{C})$ denote the matrix of $\lambda_{j,0}$, i.e.

$$J = \begin{pmatrix} & -1 \\ 1 & \end{pmatrix}.$$

**LEMMA 6.8.** *For every matrix $D \in \mathrm{M}_{2,2}(\mathbb{C})$ we have $D^t J D = \det(D)J$. If $D$ is regular we have $-JD^{-1}J = \det(D)^{-1}D^t$.*

*Proof.* Let $\lambda_{a,b}$ be the linear mapping associated with $D$, then $\lambda_{\bar{a},b}$ is the linear mapping associated with $D^t$. By Lemma 4.2 we have

$$\lambda_{\bar{a},b} \circ \lambda_{j,0} \circ \lambda_{a,b} = \lambda_{\bar{a}j,-bj} \circ \lambda_{a,b} = \lambda_{\overline{\bar{a}j}a - bj\bar{b},\, \overline{\bar{a}j}b - bj\bar{a}} = (N(a) - N(b)) \cdot \lambda_{j,0}.$$

This proves the first assertion of the lemma. The second assertion follows immediately from the first because $J^{-1} = -J$. □

**LEMMA 6.9.** *The translation planes $\mathcal{A}_{B,C}$ und $\mathcal{A}_{B',C'}$ are isomorphic if and only if the matrices $B, C$ can be transformed to the matrices $B', C'$ by a sequence of transformations of one of the following types:*

(a)
$$B' = \frac{p}{q}\frac{1}{\det(D)}D^t BD + \frac{r}{q}J$$
$$C' = \frac{p^*}{q}\frac{1}{\det(D)}D^t CD^*$$

*where $D \in \mathrm{GL}_2(\mathbb{C})$ and $p, q, r \in \mathbb{C}, p \neq 0 \neq q$.*

(b)
$$B' = -J\hat{B}J$$
$$C' = -J\hat{C}J$$

*where $\hat{C} = (C^* - B^*C^{-1}B)^{-1}$ and $\hat{B} = -C^{*-1}B^*\hat{C}^*$. If $B$ is invertible, then we also have $\hat{B} = (B - CB^{*-1}C^*)^{-1}$.*

(c)
$$B' = B^*, \quad C' = C^*$$

*Proof.* Let $\lambda : A^2 \to A^2$ be a collineation from $\mathcal{A}_{B,C}$ to $\mathcal{A}_{B',C'}$. By Theorem 1.18 we may assume that $\lambda$ is semilinear over $\mathbb{C}$.

Let $\sigma : \mathbb{C} \to \mathbb{C}$ denote the companion automorphism of $\lambda$. We extend $\sigma$ to a ring automorphism of $A$ by setting $(a + jb)^\sigma = a^\sigma + jb^\sigma$. It is easy to see that $\sigma : A \to A$ really is a ring automorphism and that $\sigma$ and $^-$ commute.

If $\mathcal{A}_{B,C}$ is non-desarguesian, then $\lambda$ fixes $S$. If $\mathcal{A}_{B,C}$ is desarguesian, then the collineation group of $\mathcal{A}_{B,C}$ acts transitively on $\mathcal{B}$, so we may assume that $\lambda$ fixes $S$ in any case. Since $\Sigma_{[s,S]}$ acts transitively on $\mathcal{B} \setminus \{S\}$, we may furthermore assume that $\lambda(W) = W$, where $W = L(0,0)$.

So there exist bijective complex linear mappings $\kappa, \mu : A \to A$ such that

$$\lambda(z, w) = (\kappa(z^\sigma), \mu^{-1}(w^\sigma))$$

for $z, w \in A$. Let $a_1, b_1, a_3, b_3 \in A$ such that

$$\kappa(z) = a_1 z + b_1 \bar{z} \quad \text{and} \quad \mu^{-1}(z) = a_3 z + b_3 \bar{z}$$

for $z \in A$. Moreover, let $D$ denote the matrix of $\mu$ and and let

$$\begin{pmatrix} p & r \\ s & q \end{pmatrix}$$

be the matrix of $\kappa$. By Lemma 6.7 (a) we then have

$$a_1 = \frac{1}{2}(p + q + j(s - r)) \quad \text{and} \quad b_1 = \frac{1}{2}(p - q + j(r + s)).$$

Let $\varphi, \psi, \varphi', \psi' : A \to A$ be defined by

$$\varphi(m) = \frac{1}{2}(m + Bm + Cm^*), \qquad \psi(m) = \frac{1}{2}(m - Bm - Cm^*),$$
$$\varphi'(m') = \frac{1}{2}(m' + B'm' + C'm'^*), \qquad \psi'(m') = \frac{1}{2}(m' - B'm' - C'm'^*).$$

It follows from Lemma 4.3 that

$$\varphi'(m') = \tfrac{1}{N(a_1)-N(b_1)}\left((a_3\varphi(m)^\sigma + b_3\overline{\psi(m)}^\sigma)\overline{a_1} - (a_3\psi(m)^\sigma + b_3\overline{\varphi(m)}^\sigma)\overline{b_1}\right),$$

$$\psi'(m') = \tfrac{1}{N(a_1)-N(b_1)}\left(-(a_3\varphi(m)^\sigma + b_3\overline{\psi(m)}^\sigma)b_1 + (a_3\psi(m)^\sigma + b_3\overline{\varphi(m)}^\sigma)a_1\right).$$

Since $\varphi'(m') + \psi'(m') = m'$, we get

$$m' = \tfrac{1}{N(a_1)-N(b_1)}\left((a_3\varphi(m)^\sigma + b_3\overline{\psi(m)}^\sigma)(\overline{a_1} - b_1)\right.$$
$$\left. + (a_3\psi(m)^\sigma + b_3\overline{\varphi(m)}^\sigma)(-\overline{b_1} + a_1)\right).$$

We distiguish the following three cases:

(a) $$s = 0, \sigma = \mathrm{id},$$

(b) $$s = 1, r = -1, p = q = 0, D = I, \sigma = \mathrm{id},$$

(c) $$p = q = 1, r = s = 0, D = I.$$

We will show that these three cases lead to the three types of transformations given in the lemma. Since every invertible $2 \times 2$-matrix is either upper triangular or it can be written as

$$\begin{pmatrix} p & r \\ & q \end{pmatrix}\begin{pmatrix} & -1 \\ 1 & \end{pmatrix}\begin{pmatrix} 1 & s \\ & 1 \end{pmatrix},$$

this will prove the lemma.

We start with case (a). Let $s = 0$, then we get $a_1 = \frac{1}{2}(p + q - jr)$ and $b_1 = \frac{1}{2}(p - q + jr)$. It follows that

$$\overline{a_1} - b_1 = \frac{1}{2}(p + q + jr - p + q - jr) = q = a_1 - \overline{b_1} \in \mathbb{C},$$
$$\overline{a_1} + b_1 = \frac{1}{2}(p + q + jr + p - q + jr) = p + jr,$$
$$a_1 + \overline{b_1} = \frac{1}{2}(p + q - jr + p - q - jr) = p - jr,$$
$$N(a_1) - N(b_1) = \det(\lambda_{a_1,b_1}) = pq.$$

Thus we get

$$m' = \frac{1}{pq}\left((a_3\varphi(m) + b_3\overline{\psi(m)})q + (a_3\psi(m) + b_3\overline{\varphi(m)})q\right)$$
$$= \frac{1}{p}(a_3 m + b_3\overline{m})$$
$$= \frac{1}{p}\mu^{-1}(m)$$

and hence

$$m = p\mu(m') = pDm'.$$

To find $B'$ and $C'$ we note that $B'm' + C'm'^* = \varphi'(m') - \psi'(m')$ for all $m' \in A$. This leads to

$$B'm' + C'm'^* = \frac{1}{pq}\left((a_3\varphi(m) + b_3\overline{\psi(m)})(\overline{a_1} + b_1)\right.$$
$$\left. + (a_3\psi(m) + b_3\overline{\varphi(m)})(-\overline{b_1} - a_1)\right)$$
$$= \frac{1}{pq}\left((a_3\varphi(m) + b_3\overline{\psi(m)})(p + jr)\right.$$
$$\left. + (a_3\psi(m) + b_3\overline{\varphi(m)})(-p + jr)\right)$$
$$= \frac{1}{pq}\left(\left(a_3(\varphi(m) - \psi(m)) - b_3\overline{(\varphi(m) - \psi(m))}\right)p\right.$$
$$\left. + (a_3 m + b_3\overline{m})jr\right)$$
$$= \frac{1}{pq}\left(\lambda_{a_3,-b_3}(\varphi(m) - \psi(m))p + m'pjr\right).$$

We now change to matrix notation. Since $D^{-1}$ is the matrix of $\mu = \lambda_{a_3,b_3}$, it follows from Lemma 4.2 that the matrix of $\lambda_{\overline{a_3},-b_3}$ is $(N(a_3) - N(b_3)) \cdot D = \det(D^{-1}) \cdot D$. By Lemma 6.7 (b) then the matrix of $\lambda_{a_3,-b_3}$ is $(\det D)^{-1} \cdot D^t$. Since $\varphi(m) - \psi(m) = Bm + Cm^*$ and and $m'j = Jm'$, we thus get

$$B'm' + C'm'^* = \frac{1}{pq}\left((\det D)^{-1}D^t(Bm + Cm^*)p) + \frac{r}{q}Jm'\right.$$
$$= \frac{1}{q}(\det D)^{-1}(D^t BpDm' + D^t Cp^* D^* m'^*) + \frac{r}{q}Jm'$$
$$= \frac{p}{q}(\det D)^{-1}D^t BDm' + \frac{p^*}{q}(\det D)^{-1}D^t CD^* m'^* + \frac{r}{q}Jm'.$$

It follows that

$$B' = \frac{p}{q}\frac{1}{\det(D)}D^t BD + \frac{r}{q}J$$
$$C' = \frac{p^*}{q}\frac{1}{\det(D)}D^t CD^*.$$

We now come to case (b). Let $s = -r = 1, p = q = 0, D = I$ and $\sigma = $ id. Then we have $a_1 = j, a_3 = 1, b_1 = b_3 = 0$. This yields

$$m' = \varphi(m)(-j) + \psi(m)j = -j(\varphi(m) - \psi(m))$$
$$= -j(Bm + Cm^*) = -j\Lambda_{B,C}(m).$$

Since $C$ is non-singular, we get by Lemma 6.3

$$m = \Lambda_{\hat{B},\hat{C}}(jm') = \Lambda_{\hat{B},\hat{C}}(Jm'),$$

where $\hat{C} = (C^* - B^*C^{-1}B)^{-1}$ and $\hat{B} = -C^{*-1}B^*\hat{C}^*$. If $B$ is invertible, then we also have $\hat{B} = (B - CB^{*-1}C^*)^{-1}$.

As in case (a) we can write

$$B'm' + C'm'^* = \varphi'(m') - \psi'(m') = \varphi(m)(-j) - \psi(m)j$$
$$= -mj = -Jm = -J(\hat{B}Jm' + \hat{C}J^*m'^*)$$
$$= -J\hat{B}Jm' - J\hat{C}Jm'^*.$$

It follows that

$$B' = -J\hat{B}J,$$
$$C' = -J\hat{C}J.$$

Finally, we treat case (c). Let $p = q = 1, r = s = 0$ and $D = I$. Then we get $a_1 = a_3 = 1$ and $b_1 = b_3 = 0$. This leads to

$$m' = \varphi(m)^\sigma + \psi(m)^\sigma = m^\sigma$$

and thus

$$m = m'^{\sigma^{-1}}.$$

As in the two other cases we have

$$B'm' + C'm'^* = \varphi'(m') - \psi'(m') = \varphi(m)^\sigma - \psi(m)^\sigma$$
$$= (Bm + Cm^*)^\sigma = B^\sigma(m'^{\sigma^{-1}})^\sigma + C^\sigma((m'^{\sigma^{-1}})^*)^\sigma$$
$$= B^\sigma m' + C^\sigma((m'^{\sigma^{-1}})^*)^\sigma.$$

This can only hold if the automorphisms $\sigma$ and $*$ commute. Then $\sigma$ fixes the field of real numbers and hence $\sigma$ is either the identity or the complex conjugation. The only interesting case is $\sigma = *$ and so we get

$$B' = B^*, \quad C' = C^*.$$

This completes the proof of the lemma. $\qquad\qquad\square$

COROLLARY 6.10. *Every collineation of $\mathcal{A}_{B,C}$ is continuous.*

Consider now the effect of a transformation of type (a) on the matrix $B$. By an appropriate choice of $r$ we can achieve that $B$ is symmetric and then we are essentially dealing with the classification of quadratic forms of a 2-dimensional complex vector space. Thus we have to distinguish the following three cases:

$(I)$
$$B = 0$$

$(II)$
$$B = \begin{pmatrix} 0 & 0 \\ 0 & 1 \end{pmatrix}$$

$(III)$
$$B = \begin{pmatrix} 0 & 0 \\ 1 & 0 \end{pmatrix}$$

The reason for choosing a non-symmetric matrix in case (III) will be become apparent during the proof of the next lemma.

LEMMA 6.11.   For every translation plane $\mathcal{A}_{B,C}$ there exists an isomorphic translation plane $\mathcal{A}_{B',C'}$ with $B' = B'^t$ and $\det(B') = 0$.

*Proof.* The lemma asserts that case (III) does not really occur. So assume that

$$B = \begin{pmatrix} 0 & 0 \\ 1 & 0 \end{pmatrix} \quad \text{and} \quad C = \begin{pmatrix} c_1 & c_2 \\ c_3 & c_4 \end{pmatrix}.$$

We first apply a transformation of type (a) which replaces $B$ by $B + s^* J$ for some $s \in \mathbb{C} \setminus \{0, -1\}$ and leaves $C$ invariant. Then we apply the transformation of type (b). This yields

$$B' = -J(\widehat{B + s^* J})J$$

where $(\widehat{B + s^* J}) = (B + s^* J - C(B^* + sJ)^{-1} C^*)^{-1}$. Here we use that $s \neq 0, -1$ and hence $B + s^* J$ is invertible. Using Lemma 6.8 we get

$$B' = -J(\widehat{B + s^* J})J$$
$$= -J(B + s^* J - C(B^* + sJ)^{-1} C^*)^{-1} J$$
$$= (\det(B + s^* J - C(B + sJ)^{-1} C^*))^{-1} \cdot (B + s^* J - C(B + sJ)^{-1} C^*)^t.$$

We claim that we can choose $s$ such that the discrimant of $B'$, i.e. the determinant of $B' + B'^t$, vanishes. Then the quadratic form defined by $B'$ becomes degenerate and the lemma is proved. The discriminant of $B'$ vanishes if and only if the discriminant of $(B + s^* J - C(B + sJ)^{-1} C^*)$ vanishes. We have

$$B + s^* J - C(B + sJ)^{-1} C^*$$

$$= \begin{pmatrix} 0 & -s^* \\ 1 + s^* & 0 \end{pmatrix} - \begin{pmatrix} c_1 & c_2 \\ c_3 & c_4 \end{pmatrix} \begin{pmatrix} 0 & -s \\ 1 + s & 0 \end{pmatrix}^{-1} \begin{pmatrix} c_1^* & c_2^* \\ c_3^* & c_4^* \end{pmatrix}$$

$$= \begin{pmatrix} 0 & -s^* \\ 1 + s^* & 0 \end{pmatrix} - \frac{1}{s(1 + s)} \begin{pmatrix} c_1 & c_2 \\ c_3 & c_4 \end{pmatrix} \begin{pmatrix} 0 & s \\ -1 - s & 0 \end{pmatrix} \begin{pmatrix} c_1^* & c_2^* \\ c_3^* & c_4^* \end{pmatrix}$$

$$= \begin{pmatrix} 0 & -s^* \\ 1 + s^* & 0 \end{pmatrix}$$

$$- \frac{1}{s(1 + s)} \begin{pmatrix} -c_2 c_1^*(1 + s) + c_1 c_3^* s & -c_2 c_2^*(1 + s) + c_1 c_4^* s \\ -c_4 c_1^*(1 + s) + c_3 c_3^* s & -c_4 c_2^*(1 + s) + c_3 c_4^* s \end{pmatrix}$$

$$= \frac{1}{s(1 + s)} \begin{pmatrix} c_2 c_1^*(1 + s) - c_1 c_3^* s & -s^* s(1 + s) + c_2 c_2^*(1 + s) - c_1 c_4^* s \\ (1 + s^*)s(1 + s) + c_4 c_1^*(1 + s) - c_3 c_3^* s & c_4 c_2^*(1 + s) - c_3 c_4^* s \end{pmatrix}$$

Let $\Delta$ denote the discriminant of this matrix, then we get

$$s^2(1 + s)^2 \Delta = 4(c_2 c_1^*(1 + s) - c_1 c_3^* s)(c_4 c_2^*(1 + s) - c_3 c_4^* s)$$
$$- (-s^* s(1 + s) + c_2 c_2^*(1 + s) - c_1 c_4^* s$$
$$+ (1 + s^*)s(1 + s) + c_4 c_1^*(1 + s) - c_3 c_3^* s)^2$$
$$= 4(c_2 c_1^*(1 + s) - c_1 c_3^* s)(c_4 c_2^*(1 + s) - c_3 c_4^* s)$$
$$- (s(1 + s) + c_2 c_2^*(1 + s) - c_1 c_4^* s + c_4 c_1^*(1 + s) - c_3 c_3^* s)^2.$$

This shows that $\Delta$ is a rational function in $s$, in particular, $\Delta$ does not depend on $s^*$. The polynomial $P(s) = s^2(1 + s)^2 \Delta$ has four roots in $\mathbb{C}$. If at least one of these roots is different from 0 and $-1$, then the lemma is proved.

So let us assume that all roots of $P(s)$ are either 0 or $-1$. At least one of these must be a multiple root. We only treat the case that 0 is a double root, the other one is similar. We have

$$P(0) = 4c_2 c_1^* c_4 c_2^* - (c_2 c_2^* + c_4 c_1^*)^2$$
$$= -(c_2 c_2^* - c_4 c_1^*)^2 = 0$$

and hence $c_2 c_2^* = c_4 c_1^* = c_1 c_4^* \in \mathbb{R}$. For the derivative of $P(s)$ we compute

$$P'(s) = 4(c_2 c_1^*(1 + s) - c_1 c_3^* s)c_4 c_2^* + 4c_2 c_1^*(c_4 c_2^*(1 + s) - c_3 c_4^* s)$$
$$- 2(s(1 + s) + c_2 c_2^*(1 + s) - c_1 c_4^* s + c_4 c_1^*(1 + s) - c_3 c_3^* s) \times$$
$$\times (s + 1 + s + c_2 c_2^* + c_4 c_1^*).$$

Thus we get

$$P'(0) = 4c_2 c_1^* c_4 c_2^* + 4c_2 c_1^* c_4 c_2^* - 2(c_2 c_2^* + c_4 c_1^*)(1 + c_2 c_2^* + c_4 c_1^*)$$
$$= 8|c_2|^4 - 4|c_2|^2(1 + 2|c_2|^2) = -4|c_2|^2 = 0.$$

It follows that $c_1 c_4^* = |c_2|^2 = 0$. But then $C$ is singular, contradicting the fact that $B$ and $C$ satisfy condition $(*)$. $\qquad\Box$

Because of Lemma 6.11 we only have to investigate the (easier) cases (I) and (II).

LEMMA 6.12. *Assume that* $B = 0$. *Then there exists* $\delta \in [0, \pi)$ *such that* $\mathcal{A}_{B,C}$ *and* $\mathcal{A}_{B',C'}$ *are isomorphic, where*

$$C' = \begin{pmatrix} 1 & \\ & e^{i\delta} \end{pmatrix}.$$

*Proof.* All transformations of type (a) with $r = 0$ leave $B = 0$ invariant.

Note that for all $z \in \mathbb{C}^2$ there holds $z^t J z = 0$. Let $(d_1, d_3)^t \in \mathbb{C}^2$ be an eigenvector of the matrix $JCJC^*$ and denote the corresponding eigenvalue by $k$. Put $(d_2, d_4)^t = -JC(d_1^*, d_3^*)^t$ and

$$D = \begin{pmatrix} d_1 & d_2 \\ d_3 & d_4 \end{pmatrix}.$$

Then we get

$$C \begin{pmatrix} d_1^* \\ d_3^* \end{pmatrix} = J \begin{pmatrix} d_2 \\ d_4 \end{pmatrix} \quad \text{and} \quad C \begin{pmatrix} d_1^* \\ d_3^* \end{pmatrix} = kJ \begin{pmatrix} d_1 \\ d_3 \end{pmatrix}.$$

It follows that $C' = D^t C D^*$ is a diagonal matrix since the off-diagonal elements of $C'$ are given by

$$\begin{pmatrix} d_2 & d_4 \end{pmatrix} C \begin{pmatrix} d_1^* \\ d_3^* \end{pmatrix} = 0 \quad \text{and} \quad \begin{pmatrix} d_1 & d_3 \end{pmatrix} C \begin{pmatrix} d_2^* \\ d_4^* \end{pmatrix} = 0.$$

So we can assume that $C$ is a diagonal matrix. We now perform a transformation of type (a) with $r = 0$ and

$$D = \begin{pmatrix} d_1 & \\ & d_4 \end{pmatrix}.$$

Then we get $B' = 0$ and

$$\begin{aligned} C' &= \frac{p^*}{q} \frac{1}{\det(D)} D^t C D^* \\ &= \frac{p^*}{q} \frac{1}{d_1 d_4} \begin{pmatrix} d_1 & \\ & d_4 \end{pmatrix} \begin{pmatrix} c_1 & \\ & c_4 \end{pmatrix} \begin{pmatrix} d_1^* & \\ & d_4^* \end{pmatrix} \\ &= \frac{p^*}{q} \frac{1}{d_1 d_4} \begin{pmatrix} d_1 c_1 d_1^* & \\ & d_4 c_4 d_4^* \end{pmatrix}. \end{aligned}$$

By a suitable choice of $p, q, d_1$ and $d_4$ we can achieve that

$$C' = \begin{pmatrix} 1 & \\ & e^{i\delta} \end{pmatrix}$$

for some $\delta \in [0, 2\pi)$. Since $B'$ and $C'$ satisfy condition $(*)$, the value $\delta = \pi$ is not allowed. An application of the transformation of type (c) replaces $\delta$ by $-\delta$, hence we can assume $\delta \in [0, \pi)$. $\qquad\square$

By calculations similar to the given ones, it can be shown that the parameter $\delta$ cannot be reduced further.

**LEMMA 6.13.** *Assume that* $B = \begin{pmatrix} 0 & 0 \\ 0 & 1 \end{pmatrix}$. *Then there exist* $r \in \mathbb{R}, c \in \mathbb{C}, r \geq 0, \operatorname{Im} c \geq 0$ *such that* $\mathcal{A}_{B,C}$ *and* $\mathcal{A}_{B,C'}$ *are isomorphic, where*

$$C' = \begin{pmatrix} 1 & r \\ & c \end{pmatrix}.$$

*Moreover, we have* $0 < P_{r,c}(x) = x^4 + (2\operatorname{Re} c - r^2)x^2 - 2rx + |c|^2 - 1$ *for all* $x \in \mathbb{R}$.

*Proof.* A short calculation shows that a transformation of type (a) leaves $B$ invariant if and only if $r = d_3 = 0$ and $pd_4^* = qd_1$, where

$$D = \begin{pmatrix} d_1 & d_2 \\ d_3 & d_4 \end{pmatrix}.$$

So assume that these conditions are satisfied, then we get

$$
\begin{aligned}
C' = \begin{pmatrix} c_1' & c_2' \\ c_3' & c_4' \end{pmatrix} &= \frac{p^*}{q} \frac{1}{\det(D)} D^t C D^* \\
&= \frac{p^*}{q} \frac{1}{d_1 d_4} \begin{pmatrix} d_1 \\ d_2 & d_4 \end{pmatrix} \begin{pmatrix} c_1 & c_2 \\ c_3 & c_4 \end{pmatrix} \begin{pmatrix} d_1^* & d_2^* \\ & d_4^* \end{pmatrix} \\
&= \frac{p^* d_1}{pd_4^*} \frac{1}{d_1 d_4} \begin{pmatrix} d_1 c_1 d_1^* & d_1 c_1 d_2^* + d_1 c_2 d_4^* \\ d_2 c_1 d_1^* + d_4 c_3 d_1^* & d_2 c_1 d_2^* + d_4 c_3 d_2^* + d_2 c_2 d_4^* + d_4 c_4 d_4^* \end{pmatrix}.
\end{aligned}
$$

By a suitable choice of the $d_j$ we can achieve $c_3' = 0$. So let us assume $c_3 = 0$. Since we also want $c_3' = 0$, we get $d_2 = 0$. This yields

$$
\begin{aligned}
c_1' &= \frac{p^*}{pd_4 d_4^*} d_1 c_1 d_1^* = \frac{p^*}{p} \frac{d_1 d_1^*}{d_4 d_4^*} c_1, \\
c_2' &= \frac{p^*}{pd_4 d_4^*} d_1 c_2 d_4^* = \frac{p^*}{p} \frac{d_1}{d_4} c_2, \\
c_4' &= \frac{p^*}{pd_4 d_4^*} d_4 c_4 d_4^* = \frac{p^*}{p} c_4.
\end{aligned}
$$

It follows that we can achieve $c_1' = 1$ and $0 \leq c_2' = r \in \mathbb{R}$. Then $c = c_4'$ cannot be reduced further.

Application of the transformation of type (c) replaces $c$ by $c^*$, hence we may assume $\operatorname{Im} c \geq 0$.

We show next that $B$ and $C'$ satisfy condition $(*)$ if and only if the polynomial $P_{r,c}(x) = x^4 + (2\operatorname{Re}c - r^2)x^2 - 2rx + |c|^2 - 1$ has no real roots. In our case condition $(*)$ says

$$|n^2| = \left| (m \ \ n) \begin{pmatrix} 0 & 0 \\ 0 & 1 \end{pmatrix} \begin{pmatrix} m \\ n \end{pmatrix} \right|$$

$$< \left| (m \ \ n) \begin{pmatrix} 1 & r \\ & c \end{pmatrix} \begin{pmatrix} m^* \\ n^* \end{pmatrix} \right| = |mm^* + mrn^* + ncn^*|$$

for all $(m, n) \in \mathbb{C}^2 \setminus \{(0,0)\}$. For $n = 0 \neq m$ this is satisfied. So assume $n \neq 0$ and put $m = zn$. This yields

$$|n|^2 < |n|^2|zz^* + zr + c|$$

for all $z \in \mathbb{C}$. Setting $z = xe^{i\vartheta}$, $x \in \mathbb{R}$, $\vartheta \in [0, 2\pi]$, we get

$$1 < |x^2 + rxe^{i\vartheta} + c|.$$

If we fix $x$ and let $\vartheta$ vary, then the right hand side of this inequality describes a circle with center $x^2 + c$ and radius $r|x|$. Hence the minimum of this right hand side is given by

$$\left| |x^2 + c| - r|x| \right|.$$

For $x = 0$ we have $1 < |c|$. Consequently we get

$$\left| |x^2 + c| - r|x| \right| = |x^2 + c| - r|x|$$

for all $x \in \mathbb{R}$ and hence our condition becomes

$$1 < |x^2 + c| - r|x|$$

for all $x \in \mathbb{R}$. This is equivalent to

$$(1 + r|x|)^2 = 1 + 2r|x| + r^2x^2$$
$$< |x^2 + c|^2 = (x^2 + \operatorname{Re}c)^2 + (\operatorname{Im}c)^2 = x^4 + (2\operatorname{Re}c)x^2 + |c|^2.$$

It follows that the polynomial $P_{r,c}(x)$ has no non-negative real roots. Of course we also have

$$1 < |x^2 + c| + r|x|$$

for all $x \in \mathbb{R}$ and this shows that the polynomial $P_{r,c}(x)$ has no negative real roots either. Hence we get the existence condition as stated in the lemma. $\quad\square$

We do not claim that the translation planes $\mathcal{A}_{B,C}$ are not isomorphic for different values of $r$ and $c$ and in fact this is not true. The parameters cannot be reduced further using transformations of type (a) and (c), but with the transformation of type (b) the situation is different. This is due to the fact that the polynomial $P(s)$ which occured in the proof of Lemma 6.11 is of degree 4 and so usually has 4 different roots. This has the consequence that for each pair of parameters $(r, c)$ there exist 3 other pairs $(r', c')$ such that the corresponding translation planes are isomorphic. In some special cases one gets $(r', c') = (r, c)$

and then the corresponding translation plane has an extra collineation. In principle, these pairs can be determined by an argument analogous to the one given in the proof of Lemma 6.11. The discriminant of the resulting matrix $B'$ is a polynomial of degree 3 in the variable $s$. For the determination of the parameters $(r', c')$ one needs explicit expressions for the roots of this polynomial. This leads to rather tedious calculations, which we decided not to perform. So our classification in case (II) is not complete, but the list in Theorem 6.6 contains at most 4 representatives from each isomorphism class.

LEMMA 6.14. *A translation plane of type* $(I_\delta)$ *is never isomorphic to a translation plane of type* $(II_{r,c})$.

*Proof.* As already mentioned in the last section, the $A_1$-indicator set $\mathcal{J}$ associated with $B$ and $C$ is normalized and hence $(A, +, \circ)$, where $m \circ z = \varphi(m)z + \psi(m)\bar{z}$ for $m, z \in A$, is a semifield which coordinatizes $\mathcal{A}_{B,C}$. The product $m \circ z, m = (m_1, m_2), z = (z_1, z_2) \in A = \mathbb{C}^2$ in the different cases is as follows:

$(I_\delta)$    $(m_1, m_2) \circ (z_1, z_2) = (m_1 z_1 - e^{i\delta} m_2^* z_2, m_1^* z_2 + m_2 z_1)$

$(II_{r,c})$   $(m_1, m_2) \circ (z_1, z_2) = (m_1 z_1 - c m_2^* z_2 - m_2 z_2, m_1^* z_2 + m_2 z_1 + r m_2^* z_2)$

In case $(I_\delta)$ the middle nucleus of $(A, +, \circ)$ consists of all $(c, 0), c \in \mathbb{C}$, whereas in case $(II_{r,c})$ it consists of all $(r, 0), r \in \mathbb{R}$. It follows now from Proposition 1.22 that translation planes of different types are not isomorphic. $\quad\square$

# 7. Locally Compact Translation Planes of Higher Dimension

In this chapter we generalize Theorem 5.1 and give a construction of locally compact 8- and 16-dimensional translation planes using contractions of the classical real division algebras $\mathbb{H}$ and $\mathbb{O}$. In contrast to Theorem 5.1 this construction does not yield all translation planes of this kind.

## 7.1 Contractions of Normed Algebras and Topological Spreads

In the sequel let $\mathbb{D}$ be one of the real alternative division algebras $\mathbb{C}$, $\mathbb{H}$ or $\mathbb{O}$ of real dimension $d = 2, 4$ or $8$, respectively. For the construction of these algebras and an exposition of their properties the reader is referred to the books by Ebbinghaus et al. [28] or Schafer [92]. The division algebra $\mathbb{D}$ admits an involutorial antiautomorphism $^- : \mathbb{D} \to \mathbb{D}$ and a positive definite quadratic norm form $N : \mathbb{D} \to \mathbb{R}$, which is given by $N(z) = z\bar{z}$ for $z \in \mathbb{D}$. The norm is multiplicative, i.e. one has $N(zw) = N(z)N(w)$ for $z, w \in \mathbb{D}$. The absolute value of $z \in \mathbb{D}$ is defined by $|z| = \sqrt{N(z)}$.

We want to investigate topological spreads of the $2d$-dimensional real vector space $\mathbb{D}^2$ which contain the subspace $S = \{0\} \times \mathbb{D}$. For $m, n \in \mathbb{D}$ we put $L(m, n) = \{(z, mz + n\bar{z}) \mid z \in \mathbb{D}\}$. The sets $L(m, n), m, n \in \mathbb{D}$, are $d$-dimensional subspaces of the real vector space $\mathbb{D}^2$ which are complementary to $S$.

**LEMMA 7.1.** Let $m_1, n_1, m_2, n_2 \in \mathbb{D}$. Then $L(m_1, n_1)$ and $L(m_2, n_2)$ are complementary if and only if $N(m_1 - m_2) \neq N(n_1 - n_2)$.

*Proof.* By construction, the dimension of $L(m_i, n_i), i = 1, 2$, as a real vector space is just one half of the dimension of $\mathbb{D}^2$. Hence, $L(m_1, n_1)$ and $L(m_2, n_2)$ are complementary if and only if their intersection is trivial.

Assume first that $L(m_1, n_1)$ and $L(m_2, n_2)$ intersect non-trivially. Then, the equation

$$m_1 z + n_1 \bar{z} = m_2 z + n_2 \bar{z}$$

has a solution $z \in \mathbb{D} \setminus \{0\}$. This equation can be rewritten as

$$(m_1 - m_2)z = (n_2 - n_1)\bar{z}.$$

Since the norm is positive definite and multiplicative, it follows that

$$N(m_1 - m_2) = N(n_1 - n_2).$$

Let now $N(m_1 - m_2) = N(n_1 - n_2)$. If $N(m_1 - m_2) = 0$, then $m_1 = m_2$ and $n_1 = n_2$, and hence $L(m_1, n_1) = L(m_2, n_2)$. So we may assume $N(m_1 - m_2) \neq 0$. In this case $n_1 - n_2$ is invertible and we have $N((n_1 - n_2)^{-1}(m_1 - m_2)) = 1$.

Since $\mathbb{D} \neq \mathbb{R}$, there exists a subalgebra $\mathbb{A}$ of $\mathbb{D}$ which is isomorphic to $\mathbb{C}$ and has the following properties. $\mathbb{A}$ contains the element $(n_1 - n_2)^{-1}(m_1 - m_2)$ and $\mathbb{A}$ is contained in the subalgebra of $\mathbb{D}$ generated by $m_1 - m_2$ and $n_1 - n_2$. By Hilbert's Satz 90 there exists an element $z \in \mathbb{A}, z \neq 0$ such that

$$-(n_1 - n_2)^{-1}(m_1 - m_2) = \frac{\bar{z}}{z}.$$

Consider now the subalgebra of $\mathbb{D}$ generated by $m_1 - m_2$ and $n_1 - n_2$. By Artin's theorem [92: III. Theorem 3.1] this subalgebra is associative. Hence the last equation can be transformed to

$$m_1 z + n_1 \bar{z} = m_2 z + n_2 \bar{z}.$$

It follows that $L(m_1, n_1)$ and $L(m_2, n_2)$ intersect non-trivially. $\qquad \square$

**THEOREM 7.2.** *Let $\mathbb{D} \neq \mathbb{R}$ be an alternative real division algebra and let $\varphi : \mathbb{D} \to \mathbb{D}$ be a mapping which satisfies the following conditions:*

(K1) *The mapping $\varphi$ is a contraction, i.e. for all $m, n \in \mathbb{D}$ with $m \neq n$ we have $|\varphi(m) - \varphi(n)| < |m - n|$.*

(K2) $\lim\limits_{|m| \to \infty} ||m| - |\varphi(m)|| = \infty.$

*Then $\mathcal{B} = \{S\} \cup \{L(m, \varphi(m)) \mid m \in \mathbb{D}\}$ is a topological spread of the real vector space $\mathbb{D}^2$.*

*Proof.* Let $\varphi$ be a mapping which satisfies (K1) and (K2) and define $\mathcal{B}$ as in the theorem. Then Lemma 7.1 implies that the elements of $\mathcal{B}$ are mutually complementary subspaces of $\mathbb{D}^2$.

We show next that every element $(z, w) \in \mathbb{D}^2$ is contained in an element of $\mathcal{B}$. The elements $(0, w)$ are contained in $S$. Hence we may assume $z \neq 0$. Let $z \neq 0$ be fixed. Then we have to show that the mapping

$$\varrho_z : \mathbb{D} \to \mathbb{D} : m \mapsto mz + \varphi(m)\bar{z}$$

is bijective.

Let $m, n \in \mathbb{D}$ with $m \neq n$. Assume that $\varrho_z(m) = \varrho_z(n)$. Then we have $(m - n)z = (\varphi(n) - \varphi(m))\bar{z}$ and since $z \neq 0$, this implies $N(m - n) = N(\varphi(m) - \varphi(n))$, contradicting (K1). Thus $\varrho_z$ is injective. From (K2) we infer that

$$|\varrho_z(m)| = |mz + \varphi(m)\bar{z}| \geq ||mz| - |\varphi(m)\bar{z}|| = |z| \cdot ||m| - |\varphi(m)|| \to \infty$$

for $|m| \to \infty$. Hence, $\varrho_z$ is proper and then also bijective by Proposition 1.25.

It remains to show that the spread $\mathcal{B}$ is topological. To this end we construct a quasifield that coordinatizes the translation plane associated with $\mathcal{B}$. We put $\varphi' = \varphi \circ \varrho_1^{-1}$ and $\psi' = \psi \circ \varrho_1^{-1}$. Further, we define a multiplication $* : \mathbb{D} \times \mathbb{D} \to \mathbb{D}$ by

$$m * z = \varphi'(m)z + \psi'(m)\bar{z}.$$

From now on we can argue exactly as in the proof of Proposition 6.1. $\qquad\square$

Theorem 7.2 gives a second proof for part of Theorem 5.1. The other assertion of Theorem 5.1 cannot be generalized since not all half-dimensional real subspaces of $\mathbb{H}^2$ and $\mathbb{O}^2$ are of the form $L(m, n)$. This raises the question if one can characterize the translation planes planes constructed in Theorem 7.2 geometrically. At present, I have no answer to this question.

As an application of our construction we prove the following

**PROPOSITION 7.3.** *Every locally compact 4-dimensional translation plane can be embedded as a Baer subplane into a locally compact 8-dimensional translation plane. The plane thus obtained can in turn be embedded as a Baer subplane into a locally compact 16-dimensional translation plane.*

*Proof.* Let $\mathcal{A}$ be a locally compact 4-dimensional translation plane. We may assume that $\mathcal{A}$ is associated with a spread $\mathcal{B}$ of the 4-dimensional real vector space $\mathbb{C}^2$ and that $S = \{0\} \times \mathbb{C} \in \mathcal{B}$. According to Theorem 5.1 then we have (up to isomorphism) $\mathcal{B} \setminus \{S\} = \{L((m, \varphi(m)) \mid m \in \mathbb{C}\}$, where $\varphi : \mathbb{C} \to \mathbb{C}$ satisfies the conditions (K1) and (K2). Let $\mathbb{D} \in \{\mathbb{H}, \mathbb{O}\}$. Then we have $\mathbb{D} = \mathbb{C} \oplus \mathbb{C}^\perp$, where the orthogonality is induced by the bilinear form associated with $N$. We define $\tilde{\varphi} : \mathbb{D} \to \mathbb{D}$ by

$$\tilde{\varphi}(m + n) = \varphi(m) \quad \text{for} \quad m \in \mathbb{C}, n \in \mathbb{C}^\perp.$$

Let $m_1, m_2 \in \mathbb{C}, n_1, n_2 \in \mathbb{C}^\perp$, then we have

$$
\begin{aligned}
|\tilde{\varphi}(m_1 + n_1) - \tilde{\varphi}(m_2 + n_2)|^2 &= |\varphi(m_1) - \varphi(m_2)|^2 \\
&\leq |m_1 - m_2|^2 \\
&\leq |m_1 - m_2|^2 + |n_1 - n_2|^2 \\
&= |m_1 + n_1 - (m_2 + n_2)|^2
\end{aligned}
$$

since $m_1 - m_2$ and $n_1 - n_2$ are perpendicular. If $m_1 + n_1 \neq m_2 + n_2$, then we can replace at least one '$\leq$'-sign by a '$<$'-sign. Hence, $\tilde{\varphi}$ satisfies (K1).

Assume that $\tilde{\varphi}$ does not satisfy (K2). Then we can find a sequence $(m_k + n_k), m_k \in \mathbb{C}, n_k \in \mathbb{C}^\perp$ such that $\lim_{k \to \infty} |m_k + n_k| = \infty$ and

$$\| |m_k + n_k| - |\tilde{\varphi}(m_k + n_k)| \| = \| |m_k + n_k| - |\varphi(m_k)| \|$$

is bounded. Since $\varphi$ satisfies (K1), we infer from

$$\| |m_k + n_k| - |\varphi(m_k)| \| \geq \| |m_k| - |\varphi(m_k)| \|$$

that the sequence $(m_k)$ is bounded as well. But then $(n_k)$ is also bounded, contradicting the fact that $(m_k + n_k)$ is unbounded. Hence, $\tilde{\varphi}$ also satisfies (K2) and by Theorem 7.2 $\tilde{\varphi}$ generates an 8- or 16-dimensional locally compact translation plane.

By construction, the plane for $\mathbb{D} = \mathbb{H}$ is a subplane of the plane for $\mathbb{D} = \mathbb{O}$. Furthermore, the plane we started with is a subplane of both of these planes. By general results of Salzmann [89: 1.4; 90: 1.4], a $d$-dimensional subplane of a $2d$-dimensional plane is necessarily a Baer subplane. $\qquad\square$

In [59], Hiramine, Matsumoto and Oyama show that from every spread of a 4-dimensional vector space over a finite field of odd order one can construct a spread of an 8-dimensional vector space over the same field. However, in general their construction does not yield an embedding of the original plane into the newly constructed plane.

## 7.2 Locally Compact Translation Planes with Large Collineation Group

In this section we study mappings which satisfy (K1) and (K2) and have certain symmetry properties, e.g. they do not really depend on $z$ but only on the real part or only on the absolute value of $z$. The corresponding translation planes then admit certain large collineation groups. In fact, it follows from Hähl's classification of the locally compact 16-dimensional translation planes with an at least 38-dimensional collineation group that all these planes can be described by our methods.

Let $\mathbb{D}$ be an alternative real division algebra. For $m \in \mathbb{D}$ we define the *real part* by $\operatorname{Re} m = \frac{1}{2}(m + \overline{m})$ and the *pure part* by $\operatorname{Pu} m = \frac{1}{2}(m - \overline{m})$. Furthermore, we put $\operatorname{Pu}\mathbb{D} = \{m \in \mathbb{D} \mid m + \overline{m} = 0\} = \{m \in \mathbb{D} \mid m^2 \in (-\infty, 0]\}$ and $\mathbb{S}\operatorname{Pu}\mathbb{D} = \{m \in \operatorname{Pu}\mathbb{D} \mid N(m) = 1\} = \{m \in \mathbb{D} \mid m^2 = -1\}$.

Let $\operatorname{Aut}\mathbb{D}$ denote the group of all real linear automorphisms of $\mathbb{D}$. The structure of $\operatorname{Aut}\mathbb{D}$ is well-known: the only real linear automorphisms of $\mathbb{C}$ are the identity and the conjugation, $\mathbb{H}$ is a finite-dimensional central simple associative algebra over $\mathbb{R}$ and hence $\mathbb{H}$ has only inner automorphisms by the Skolem-Noether theorem [63: p.222], and the automorphism group of $\mathbb{O}$ is the 14-dimensional compact simple Lie group $G_2$. All elements of $\operatorname{Aut}\mathbb{D}$ leave the norm invariant und commute with the canonical antiautomorphism $^{-}$.

PROPOSITION 7.4. *Let* $\mathbb{D} \in \{\mathbb{C}, \mathbb{H}, \mathbb{O}\}$ *and let* $\psi : \mathbb{R} \to \mathbb{R}$ *be a mapping for which the conditions (K1) and (K2) from Theorem 7.2 are satisfied. Furthermore, let* $\mu \in \mathbb{R}$ *with* $|\mu| < 1$. *We define a mapping* $\varphi : \mathbb{D} \to \mathbb{D}$ *by*

$$\varphi(m) = \psi(\operatorname{Re} m) + \mu\operatorname{Pu} m.$$

Then $\varphi$ also satisfies (K1) and (K2). Let $A_{\varphi,\mu}$ denote the translation plane associated with $\varphi$ and $\mu$. Then $A_{\varphi,\mu}$ admits the following collineations:

$$\vartheta_\alpha : (z,w) \mapsto (\alpha(z), \alpha(w)), \quad \alpha \in \mathrm{Aut}\,\mathbb{D},$$

$$\sigma_p : (z,w) \mapsto (z, w + pz + \mu p\overline{z}), \quad p \in \mathrm{Pu}\,\mathbb{D},$$

$$\kappa_r : (z,w) \mapsto (zr, wr), \quad r \in \mathbb{R}^\times.$$

In case $\psi(tx) = t\psi(x)$ for all $t, x \in \mathbb{R}, t > 0$, the mappings

$$\lambda_t : (z,w) \mapsto (z, wt), \quad t \in \mathbb{R}, t > 0,$$

are also collineations. If $\psi$ is real linear the mappings

$$\tau_s : (z,w) \mapsto (z, w + sz + \psi(s)\overline{z}), \quad s \in \mathbb{R},$$

$$\beta : (z,w) \mapsto (z, -w)$$

are collineations as well.

*Proof.* Assume that $\psi : \mathbb{R} \to \mathbb{R}$ satisfies (K1) and (K2) and let $\mu \in \mathbb{R}$ with $|\mu| < 1$. Define $\varphi : \mathbb{D} \to \mathbb{D}$ as in the proposition.

Let $m, n \in \mathbb{D}$, then we have

$$\begin{aligned}
|\varphi(m) - \varphi(n)|^2 &= |\psi(\mathrm{Re}\,m) - \psi(\mathrm{Re}\,n) + \mu\mathrm{Pu}\,(m-n)|^2 \\
&= |\psi(\mathrm{Re}\,m) - \psi(\mathrm{Re}\,n)|^2 + |\mu|^2|\mathrm{Pu}\,(m-n)|^2 \\
&\le |\mathrm{Re}\,(m-n)|^2 + |\mu|^2|\mathrm{Pu}\,(m-n)|^2 \\
&\le |m-n|^2.
\end{aligned}$$

Just as in the proof of Proposition 7.3 this implies that $\varphi$ satisfies (K1).

The validity of (K2) also follows as in the proof of Proposition 7.3.

We show next that the mappings given in the proposition are collineations of $A_{\varphi,\mu}$.

Let $\alpha \in \mathrm{Aut}\,\mathbb{D}$ and $m \in \mathbb{D}$, then we have

$$\begin{aligned}
\vartheta_\alpha(z, mz + \varphi(m)\overline{z}) &= (\alpha(z), \alpha(m)\alpha(z) + \alpha(\varphi(m))\overline{\alpha(z)}) \\
&= (\alpha(z), \alpha(m)\alpha(z) + \alpha(\psi(\mathrm{Re}\,m) + \mu\mathrm{Pu}\,m)\overline{\alpha(z)}) \\
&= (\alpha(z), \alpha(m)\alpha(z) + (\psi(\mathrm{Re}\,m) + \mu\alpha(\mathrm{Pu}\,m))\overline{\alpha(z)}) \\
&= (\alpha(z), \alpha(m)\alpha(z) + \varphi(\alpha(m))\overline{\alpha(z)}).
\end{aligned}$$

Hence, $\vartheta_\alpha$ is a collineation of $A_{\varphi,\mu}$.

Let $p \in \mathrm{Pu}\,\mathbb{D}$ and $m \in \mathbb{D}$, then we have

$$\begin{aligned}
\sigma_p(z, mz + \varphi(m)\overline{z}) &= (z, mz + \varphi(m)\overline{z} + pz + \mu p\overline{z}) \\
&= (z, (m+p)z + (\psi(\mathrm{Re}\,m) + \mu(\mathrm{Pu}\,m + p))\overline{z}) \\
&= (z, (m+p)z + \varphi(m+p)\overline{z})
\end{aligned}$$

since $\mathrm{Re}\,(m+p) = \mathrm{Re}\,m$. Thus, $\sigma_p$ is a collineation of $A_{\varphi,\mu}$.

The mappings $\kappa_r, r \in \mathbb{R}^{\times}$ are kernel homologies and hence collineations of $\mathcal{A}_{\varphi,\mu}$.

Assume now that $\psi(tx) = t\psi(x)$ for all $x, t \in \mathbb{R}, t > 0$. Let $t > 0$ and $m \in \mathbb{D}$, then we have

$$\begin{aligned}
\lambda_t(z, mz + \varphi(m)\bar{z}) &= (z, tmz + t\varphi(m)\bar{z}) \\
&= (z, tmz + (t\psi(\operatorname{Re} m) + t\mu\operatorname{Pu} m)\bar{z}) \\
&= (z, tmz + (\psi(\operatorname{Re}(tm)) + \mu\operatorname{Pu}(tm))\bar{z}) \\
&= (z, tmz + \varphi(tm)\bar{z}).
\end{aligned}$$

Hence, $\lambda_t$ is a collineation of $\mathcal{A}_{\varphi,\mu}$.

If $\psi$ is linear, a similar reasoning shows that then the mappings $\tau_s, s \in \mathbb{R}$, and $\beta$ are also collineations of $\mathcal{A}_{\varphi,\mu}$. $\qquad\square$

In case $\mathbb{D} = \mathbb{H}$ the collineation group of $\mathcal{A}_{\varphi,\mu}$ contains a subgroup isomorphic to $SO_3(\mathbb{R})$ and a 3-dimensional group of shears with axis $S$ and center $s$. The 8-dimensional translation planes with these properties were determined by Hähl [46] and it can be shown that they are precisely the planes constructed in Proposition 7.4. In case $\mathbb{D} = \mathbb{O}$ the collineation group of $\mathcal{A}_{\varphi,\mu}$ contains a subgroup isomorphic to $G_2$ and a 7-dimensional group of shears with axis $S$ and center $s$. Again, the 16-dimensional translation planes with these properties were determined by Hähl [49] and they are precisely the planes from Proposition 7.4.

PROPOSITION 7.5. *Let $\mathbb{D} \in \{\mathbb{C}, \mathbb{H}, \mathbb{O}\}$ and let $\psi : [0, \infty) \to \mathbb{R}$ be a function with $\psi(0) = 0$ which satisfies (K1) and (K2). We define a mapping $\varphi : \mathbb{D} \to \mathbb{D}$ by*

$$\varphi(m) = \begin{cases} \frac{m}{|m|}\psi(|m|) & \text{if } m \neq 0 \\ 0 & \text{if } m = 0. \end{cases}$$

*Then $\varphi$ satisfies (K1) and (K2). Let $\mathcal{A}_{\varphi}$ denote the translation plane associated with $\varphi$. Then $\mathcal{A}_{\varphi}$ admits the following collineations:*

$$\kappa_r : (z, w) \mapsto (zr, wr), \quad r \in \mathbb{R}^{\times}.$$

*In case $\mathbb{D} = \mathbb{C}$ the mappings*

$$\lambda_c : (z, w) \mapsto (z, cw), \quad c \in \mathbb{C}, |c| = 1,$$

*are also collineations of $\mathcal{A}_{\varphi}$.*
*In case $\mathbb{D} = \mathbb{H}$ the mappings*

$$\vartheta_{a,b} : (z, w) \mapsto (bzb^{-1}, awb^{-1}), \quad a, b \in \mathbb{H}, |a| = |b| = 1,$$

*are also collineations of $\mathcal{A}_{\varphi}$. These mappings constitute a group isomorphic to $SO_4(\mathbb{R})$.*
*In case $\mathbb{D} = \mathbb{O}$ the mappings*

$$\sigma_c : (z, w) \mapsto (-czc, wc), \quad c \in \mathbb{SPu}\,\mathbb{O},$$

*are also collineations of $\mathcal{A}_\varphi$. These mappings generate a group isomorphic to* Spin$_7$.

*If there exists $\beta \in [0, \infty)$ such that $\varphi\left(\frac{1}{2}(t + t^\beta)\right) = \frac{1}{2}(t - t^\beta)$ for all $t \in [0, \infty)$, then the mappings*

$$\xi_t : (z, w) \mapsto (tz + \psi(t)\bar{z}, w), \quad t \in (0, \infty),$$

*are also collineations of $\mathcal{A}_\varphi$.*

*Proof.* Let $\psi : [0, \infty) \to \mathbb{R}$ be a function with $\psi(0) = 0$ which satisfies (K1) and (K2) and let $\varphi : \mathbb{D} \to \mathbb{D}$ be defined as in the proposition.

Let $m, n \in \mathbb{D} \setminus \{0\}$ with $m \neq n$. By Artin's theorem, the subalgebra of $\mathbb{D}$ generated by $m$ and $n$ is associative. Hence we have

$$|\varphi(m) - \varphi(n)| = \left| \frac{m}{|m|} \psi(|m|) - \frac{n}{|n|} \psi(|n|) \right| = \left| \psi(|m|) - m^{-1}n|mn^{-1}|\psi(|n|) \right|.$$

We choose an element $p \in \mathbb{S}\mathrm{Pu}\mathbb{D}$ such that $m^{-1}n|mn^{-1}|$ is contained in the subalgebra of $\mathbb{D}$ generated by $p$. This subalgebra is isomorphic to $\mathbb{C}$. Since $|m^{-1}n|mn^{-1}|| = 1$, we therefore find an element $t \in \mathbb{R}$ with $m^{-1}n|mn^{-1}| = e^{pt}$. As $m \neq n$, we cannot have $|m| = |n|$ and $t = 2k\pi, k \in \mathbb{Z}$, at the same time. Moreover, there holds

$$|m - n| = ||m| - m^{-1}n|mn^{-1}| \cdot |n|| = ||m| - e^{pt}|n||.$$

In order to prove (K1) it is therefore sufficient to show that

$$|\psi(r) - e^{pt}\psi(s)| < |r - e^{pt}s|$$

for all $r, s \in (0, \infty), p \in \mathbb{S}\mathrm{Pu}\mathbb{D}, t \in \mathbb{R}$ with $r \neq s$ or $t \notin 2\pi\mathbb{Z}$. Squaring both sides of this inequality gives us the equivalent inequality

$$\psi(r)^2 + \psi(s)^2 - 2\psi(r)\psi(s)\cos t < r^2 + s^2 - 2rs\cos t,$$

or rewritten

$$f(t) = \psi(r)^2 + \psi(s)^2 - 2\psi(r)\psi(s)\cos t - r^2 - s^2 + 2rs\cos t < 0.$$

If the function $f$ assumes its maximum in the point $t \in \mathbb{R}$, then

$$f'(t) = 2\sin t(\psi(r)\psi(s) - rs) = 0.$$

Since $\psi(0) = 0$ and $\psi$ satisfies (K1), we have $\psi(r) < r$ for $r > 0$. Hence, the zeros of $f'$ are precisely the points $k\pi, k \in \mathbb{Z}$. We have $f(k\pi) = (\psi(r) + (-1)^k\psi(s))^2 - (r + (-1)^k s)^2$ for $k \in \mathbb{Z}$. Thus $f(t) = 0$ if and only if $r = s$ and $t = 2k\pi, k \in \mathbb{Z}$. It follows that $\varphi$ satisfies (K1).

Let $m \in \mathbb{D}$, then we have

$$||m| - |\varphi(m)|| = ||m| - |\psi(|m|)||.$$

Hence, (K2) carries over from $\psi$ to $\varphi$.

' The mappings $\kappa_r$, $r \in \mathbb{R}^\times$, are kernel homologies and hence they are collineations of $\mathcal{A}_\varphi$.

Let $\mathbb{D} = \mathbb{H}$ and let $a, b \in \mathbb{H}$ with $|a| = |b| = 1$, then we have

$$\vartheta_{a,b}(z, mz + \varphi(m)\bar{z}) = (bzb^{-1}, amzb^{-1} + a\frac{m}{|m|}\psi(|m|)\bar{z}b^{-1})$$

$$= (bzb^{-1}, amb^{-1}bzb^{-1} + \frac{amb^{-1}}{|amb^{-1}|}\psi(|amb^{-1}|)\overline{bzb^{-1}}).$$

Hence $\vartheta_{a,b}$ is a collineation of $\mathcal{A}_\varphi$. A similar argument applies to the case $\mathbb{D} = \mathbb{C}$.

Let $\mathbb{D} = \mathbb{O}$ and $c \in \mathbb{S}\mathrm{Pu}\,\mathbb{O}$. Then we have $\bar{c} = -c$ and $c^2 = -\bar{c}c = -1$. Let $m, z \in \mathbb{O}$. Then the Moufang identities imply $(mc)(-czc) = (mz)c$. This argument is adopted from [48: Lemma 3.1]. Thus we get

$$\sigma_c(z, mz + \varphi(m)\bar{z}) = (-czc, (mz)c + (\varphi(m)\bar{z})c)$$

$$= (-czc, (mc)(-czc) + (\varphi(m)c)(-c\bar{z}c))$$

$$= (-czc, (mc)(-czc) + \frac{mc}{|m|}\psi(|m|)\overline{-czc})$$

$$= (-czc, (mc)(-czc) + \varphi(mc)\overline{-czc}).$$

Hence $\sigma_c$ is a collineation of $\mathcal{A}_\varphi$. In [48: Lemma 3.1] it is shown that the group generated by $\{\sigma_c \mid c \in \mathbb{S}\mathrm{Pu}\,\mathbb{O}\}$ is isomorphic to $\mathrm{Spin}(7)$.

Assume now that there exists $\beta \in [0, \infty)$ such that $\psi\left(\frac{1}{2}(t + t^\beta)\right) = \frac{1}{2}(t - t^\beta)$ for all $t \in [0, \infty)$. Let $x \in [0, \infty)$, then there exists precisely one $t \in [0, \infty)$ with $x = \frac{1}{2}(t + t^\beta)$. Since $\psi(x) = \frac{1}{2}(t - t^\beta)$, it follows that

$$(*) \qquad\qquad (x + \psi(x))^\beta = x - \psi(x)$$

for all $x \in [0, \infty)$.

Let $t \in (0, \infty)$ and $z, m \in \mathbb{D}$, then we have

$$\xi_t(z, mz + \varphi(m)\bar{z}) = (tz + \psi(t)\bar{z}, mz + \varphi(m)\bar{z})$$

$$= \left(w, m\left(\frac{tw - \psi(t)\overline{w}}{t^2 - \psi(t)^2}\right) + \varphi(m)\left(\overline{\frac{tw - \psi(t)\overline{w}}{t^2 - \psi(t)^2}}\right)\right)$$

$$= \left(w, \frac{1}{t^2 - \psi(t)^2}((mt - \varphi(m)\psi(t))w + (\varphi(m)t - m\psi(t))\overline{w})\right),$$

where $w = tz + \psi(t)\bar{z}$ and accordingly $z = \frac{tw - \psi(t)\overline{w}}{t^2 - \psi(t)^2}$. In order to see that $\xi_t$ is a collineation of $\mathcal{A}_\varphi$ it thus remains to show

$$\varphi\left(\frac{1}{t^2 - \psi(t)^2}(mt - \varphi(m)\psi(t))\right) = \frac{1}{t^2 - \psi(t)^2}(\varphi(m)t - m\psi(t))$$

for all $m \in \mathbb{D}, t \in (0, \infty)$. This is obvious for $m = 0$, hence we may assume $m \neq 0$. Then we get

$$\varphi\left(\frac{1}{t^2 - \psi(t)^2}(mt - \varphi(m)\psi(t))\right) = \frac{m}{|m|} \cdot \psi\left(\left|\frac{t|m| - \psi(t)\psi(|m|)}{t^2 - \psi(t)^2}\right|\right)$$

and

$$\frac{1}{t^2 - \psi(t)^2}(\varphi(m)t - m\psi(t)) = \frac{m}{|m|} \cdot \frac{t\psi(|m|) - |m|\psi(t)}{t^2 - \psi(t)^2}.$$

From $(*)$ we infer

$$\left(\frac{|m| + \psi(|m|)}{t + \psi(t)}\right)^{\beta} = \frac{|m| - \psi(|m|)}{t - \psi(t)}.$$

Thus we can calculate as follows

$$\psi\left(\left|\frac{t|m| - \psi(t)\psi(|m|)}{t^2 - \psi(t)^2}\right|\right) = \psi\left(\frac{t|m| - \psi(t)\psi(|m|)}{t^2 - \psi(t)^2}\right)$$

$$= \psi\left(\frac{1}{2}\frac{|m| + \psi(|m|)}{t + \psi(t)} + \frac{1}{2}\frac{|m| - \psi(|m|)}{t - \psi(t)}\right)$$

$$= \psi\left(\frac{1}{2}x + \frac{1}{2}x^{\beta}\right) = \frac{1}{2}x - \frac{1}{2}x^{\beta}$$

$$= \frac{1}{2}\frac{|m| + \psi(|m|)}{t + \psi(t)} - \frac{1}{2}\frac{|m| - \psi(|m|)}{t - \psi(t)}$$

$$= \frac{t\psi(|m|) - |m|\psi(t)}{t^2 - \psi(t)^2}.$$

So $\xi_t$ is a collineation of $\mathcal{A}_{\varphi}$. $\qquad\square$

The locally compact 8- and 16-dimensional translation planes whose collineation group contains a subgroup isomorphic to $SO_4(\mathbb{R})$ or $Spin_7$, respectively, were determined by Hähl [44; 48]. It can be shown that they are precisely the planes $\mathcal{A}_{\varphi}$ from Proposition 7.5.

It follows from unpublished results of Hähl that every locally compact 16-dimensional translation plane with an at least 38-dimensional collineation group is isomorphic to one of the planes constructed in Proposition 7.4 or 7.5, cf. [42: p.264].

# Generalizations and Extensions

In this final section we discuss some generalizations and extensions of the methods developed up to now.

Let $L$ be a Galois extension of a commutative field $F$ with Galois group $\mathcal{G}$. By Dedekind's lemma [62: 4.14] the elements of $\mathcal{G}$ are linearly independent over $L$. Hence, every $F$-linear mapping $\lambda : L \to L$ can be written as

$$\lambda : z \mapsto \sum_{\sigma \in \mathcal{G}} a_\sigma \cdot \sigma(z),$$

where the elements $a_\sigma \in L$ are uniquely determined. This fact was used by R. Liebler [70] for the investigation of finite planes of Lenz type V. In order to generalize Theorem 2.6 to this situation it is necessary to express the determinant of $\lambda$ as a function of the $a_\sigma, \sigma \in \mathcal{G}$.

Let $b_1, \ldots, b_4 \in \mathbb{H}$ be linearly independent over $\mathbb{R}$. Then every $\mathbb{R}$-linear mapping $\lambda : \mathbb{H} \to \mathbb{H}$ can be expressed as

$$\lambda : z \mapsto \sum_{i=1}^{4} a_i z b_i,$$

where $a_1, \ldots, a_4 \in \mathbb{H}$ are uniquely determined. This fact was already known to Hamilton, cf. [28: 6.§1.6]. The determinant of $\lambda$ as a function of $a_1, \ldots, a_4$, $b_1, \ldots, b_4$ was computed by Schrutka von Rechtenstamm [95: (210)].

The analogous result for the representation of $\mathbb{R}$-linear mappings $\lambda : \mathbb{O} \to \mathbb{O}$ was obtained by Greub [35]. Let $b_1, \ldots, b_8 \in \mathbb{O}$ be linearly independent over $\mathbb{R}$. Then $\lambda$ can be written as

$$\lambda : z \mapsto \sum_{i=1}^{8} (a_i z) b_i,$$

where $a_1, \ldots, a_8 \in \mathbb{O}$ are uniquely determined. The determinant of $\lambda$ as a function of $a_1, \ldots, a_8, b_1, \ldots, b_8$ has not yet been computed.

# Bibliography

[1] André, J.: Über nicht-Desarguessche Ebenen mit transitiver Translations-gruppe. Math. Z. **60**, 156–186 (1954)

[2] André, J.: Projektive Ebenen über Fastkörpern. Math. Z. **62**, 137–160 (1955)

[3] Bachmann, F.: Aufbau der Geometrie aus dem Spiegelungsbegriff. 2. Auflage. Berlin-Heidelberg-New York. Springer 1973.

[4] Bader, L. and Lunardon, G.: On the flocks of $Q^+(3, q)$. Geom. Dedicata **29**, 177–183 (1989)

[5] Benz, W.: Vorlesungen über Geometrie der Algebren. Berlin-Heidelberg-New York. Springer 1973.

[6] Bernardi, M.: Esistenzia di fibrazioni in uno spazio proiettivo infinito. Istit. Lombardo Accad. Sci. Lett. Rend.A **107**, 528–542 (1973)

[7] Bernardi, M. et Torre, A.: Problemi di estensione per fibrazioni proiettive. Boll. U. M. I. (5) **18-B**, 989–1002 (1981)

[8] Betten, D.: 4-dimensionale Translationsebenen. Math. Z. **128**, 129–151 (1972)

[9] Betten, D.: 4-dimensionale Translationsebenen mit 8-dimensionaler Kollineationsgruppe. Geom. Dedicata **2**, 327–339 (1973)

[10] Betten, D.: 4-dimensionale Translationsebenen mit irreduzibler Kollineationsgruppe. Arch. Math. **24**, 552–560 (1973)

[11] Betten, D.: 4-dimensionale Translationsebenen mit genau einer Fixrichtung. Geom. Dedicata **3**, 405–440 (1976)

[12] Betten, D.: 4-dimensionale Translationsebenen mit 7-dimensionaler Kollineationsgruppe. J. Reine Angew. Math. **285**, 126–148 (1976)

[13] Betten, D.: 4-dimensionale Translationsebenen mit kommutativer Standgruppe. Math. Z. **154**, 125–141 (1977)

[14] Blaschke, W.: Projektive Geometrie. Wolfenbüttel und Hannover 1948.

[15] Brauner, H.: Geometrie projektiver Räume II. Mannheim-Wien-Zürich. BI 1976.

[16] Breuning, P.: Translationsebenen und Vektorraumbündel. Mitt. Math. Sem. Gießen 86, 1–50 (1970)

[17] Bröcker, L.: Kinematische Räume. Geom. Dedicata 1, 241–278 (1973)

[18] Bruck, R. H. and Bose, R. C.: The construction of translation planes from projective spaces. J. Alg. 1, 85–102 (1964)

[19] Bruck, R. H. and Bose, R. C.: Linear representations of projective planes in projective spaces. J. Alg. 4, 117–172 (1966)

[20] Bruen, A.: Spreads and a conjecture of Bruck and Bose. J. Alg. 23, 519–537 (1972)

[21] Bruen, A. and Fisher, J. C.: Spreads which are not dual spreads. Can. Math. Bull. 12, 801–803 (1969)

[22] Buchanan, T. and Hähl, H.: On the kernel and the nuclei of 8-dimensional locally compact quasifields. Arch. Math. 24, 472–480 (1977)

[23] Buchanan, T. and Hähl, H.: The transposition of locally compact connected translation planes. J. Geom. 11, 84–92 (1978)

[24] Burau, W.: Mehrdimensionale projektive und höhere Geometrie. Berlin. VEB Dt. Verlag der Wissenschaften 1961.

[25] Cohn, P. M.: Skew field constructions. Cambridge. Cambridge University Press 1977.

[26] Dembowski, P.: Finite geometries. Berlin-Heidelberg-New York. Springer 1968.

[27] Dugundji, J.: Topology. Boston, Massachusetts. Allyn and Bacon 1966.

[28] Ebbinghaus, H.-D. et al.: Numbers. Heidelberg-Berlin-New York-Tokyo. Springer 1990.

[29] Eilenberg, S. and Steenrod, N.: Foundations of algebraic topology. Princeton. Princeton University Press 1952.

[30] Fisher, J. C.: Geometry according to Euclid. Am. Math. Monthly 86, 260–270 (1979)

[31] Fisher, J. C. and Thas, J. A.: Flocks in PG(3,q). Math. Z. 169, 1–11 (1979)

[32] Gevaert,H. and Johnson, N. L.: Flocks of quadratic cones, generalized quadrangles and translation planes. Geom. Dedicata 27, 301–317 (1988)

[33] Gevaert, H., Johnson, N. L. and Thas, J. A.: Spreads covered by reguli. Simon Stevin 62, 51–62 (1988)

[34] Gluck, H. and Warner, F. W.: Great circle fibrations of the three-sphere. Duke Math. J. **50**, 107–132 (1983)

[35] Greub, W. H.: The Cayley algebra and linear transformations of $\mathbb{R}^8$. C. R. Math. Acad. Sci., Soc. R. Can. **2**, 135–140 (1980)

[36] Grundhöfer, Th.: Reguli in Faserungen projektiver Räume. Geom. Dedicata **11**, 227–237 (1981)

[37] Grundhöfer, Th. and Salzmann, H.: Locally compact double loops and ternary fields. Chapter XI of Quasigroups and loops — Theory and applications. Chein, O., Pflugfelder, H. and Smith, J. H. D. (eds.). Berlin. Heldermann 1990.

[38] Hähl, H.: Automorphismengruppen von lokalkompakten zusammenhängenden Quasikörpern und Translationsebenen. Geom. Dedicata **4**, 305–321 (1975)

[39] Hähl, H.: Vierdimensionale reelle Divisionsalgebren mit dreidimensionaler Automorphismengruppe. Geom. Dedicata **4**, 323–331 (1975)

[40] Hähl, H.: Geometrisch homogene vierdimensionale reelle Divisionsalgebren. Geom. Dedicata **4**, 333–361 (1975)

[41] Hähl, H.: Automorphismengruppen achtdimensionaler lokalkompakter Quasikörper. Math. Z. **149**, 203–225 (1976)

[42] Hähl, H.: Zur Klassifikation von 8- und 16-dimensionalen lokalkompakten Translationsebenen nach ihren Kollineationsgruppen. Math. Z. **159**, 259–294 (1978)

[43] Hähl, H.: Achtdimensionale lokalkompakte Translationsebenen mit großen Streckungsgruppen. Arch. Math. **34**, 231–242 (1980)

[44] Hähl, H.: Achtdimensionale lokalkompakte Translationsebenen mit großen kompakten Kollineationsgruppen. Mh. Math. **90**, 207–218 (1980)

[45] Hähl, H.: Kriterien für lokalkompakte topologische Quasikörper. Arch. Math. **38**, 273–279 (1982)

[46] Hähl, H.: Eine Klasse von achtdimensionalen lokalkompakten Translationsebenen mit großen Scherungsgruppen. Mh. Math. **97**, 23–45 (1984)

[47] Hähl, H.: Achtdimensionale lokalkompakte Translationsebenen mit mindestens 17-dimensionaler Kollineationsgruppe. Geom. Dedicata **21**, 299–340 (1986)

[48] Hähl, H.: Sechzehndimensionale lokalkompakte Translationsebenen mit Spin(7) als Kollineationsgruppe. Arch. Math. **48**, 267–276 (1987)

[49] Hähl, H.: Sechzehndimensionale lokalkompakte Translationsebenen, deren Kollineationsgruppe $G_2$ enthält. Geom. Dedicata **36**, 181–197 (1990)

[50] Havlicek, H.: Zur Theorie linearer Abbildungen I. J. Geom. **16**, 152–167 (1981)

[51] Havlicek, H.: Zur Theorie linearer Abbildungen II. J. Geom. **16**, 168–180 (1981)

[52] Havlicek, H.: Dual spreads generated by collineations. Simon Stevin **64**, 339–349 (1990)

[53] Havlicek, H.: On sets of lines corresponding to affine spaces. Proc. Combinatorics 88. Vol.1, p. 449–457

[54] Heimbeck, G.: Regelflächenscharen. Geom. Dedicata **22**, 235–245 (1987)

[55] Heimbeck, G. und Wagner, R.: Eine neue Serie von endlichen Translationsebenen. Geom. Dedicata **28**, 107–125 (1988)

[56] Herzer, A.: Eine Verallgemeinerung des Satzes von Dandelin. Elem. Math. **27**, 52–56 (1972)

[57] Herzer, A.: Charakterisierung regulärer Faserungen durch Schließungssätze. Arch. Math. **25**, 662–672 (1974)

[58] Herzer, A. und Lunardon, G.: Charakterisierung (A,B)-regulärer Faserungen durch Schließungssätze. Geom. Dedicata **6**, 471–484 (1977)

[59] Hiramine, Y., Matsumoto, M. and Oyama, T.: On some extension of 1-spread sets. Osaka J. Math. **24**, 123–137 (1987)

[60] Hocking, J. G. and Young, G. S.: Topology. Reading. Addison-Wesley 1961.

[61] Hughes, D. R. and Piper, F. C.: Projective planes. Berlin-Heidelberg-New York. Springer 1973.

[62] Jacobson, N.: Basic algebra I. 2nd edition. New York. Freeman 1985.

[63] Jacobson, N.: Basic algebra II. New York. Freeman 1980.

[64] Johnson, N. L.: Flocks of hyperbolic quadrics and translation planes admitting affine homologies. J. Geom. **34**, 50–73 (1989)

[65] Karzel, H. und Kist, G.: Zur Begründung metrischer affiner Ebenen. Abh. Math. Sem. Univ. Hamb. **49**, 234–236 (1979)

[66] Karzel, H. and Kist, G.: Kinematic algebras and their geometries. In: Rings and Geometry. ed. by R. Kaya, P. Plaumann and K. Strambach. Dordrecht. Reidel 1985. p. 437–509

[67] Kestelmann, H.: Automorphisms of the field of complex numbers. Proc. London Math. Soc. (2) **53**, 1–12 (1951)

[68] Kolb, E.: The Schwan/Artin coordinatization for nearfield planes. Geom. Dedicata **50**, 283–290 (1994)

[69] Kühne, R. and Löwen. R.: Topological projective spaces. Abh. Math. Sem. Univ. Hamburg **62**, 1–9 (1992)

[70] Liebler, R. A.: On nonsingular tensors and related projective planes. Geom. Dedicata **11**, 455–464 (1981)

[71] Löwen, R.: Compact spreads and compact translation planes over locally compact fields. J. Geom. **36**, 110–116 (1989)

[72] Löwen, R.: Topological pseudo-ovals, elation Laguerre planes, and generalized quadrangles. Math. Z. **216**, 347–369 (1994)

[73] Lüneburg, H.: Einige methodische Bemerkungen zur Theorie der elliptischen Ebenen. Abh. Math. Sem. Univ. Hamb. **34**, 59–72 (1970)

[74] Lüneburg, H.: Translation planes. Berlin-Heidelberg-New York. Springer 1980.

[75] Lunardon, G.: Proposizioni configurazionali in una classe di fibrazioni. Boll. U. M. I. (5) **13-A**, 404–413 (1976)

[76] Lunardon,G.: Insiemi indicatori proiettivi e fibrazioni planari di uno spazio proiettivo finito. Boll. U. M. I. (6) **3-B**, 717–735 (1984)

[77] Maduram, D. M.: Transposed translation planes. Proc. Am. Math. Soc. **53**, 265–270 (1975)

[78] Maduram, D. M.: Matrix representation of translation planes. Geom. Dedicata **4**, 485–492 (1975)

[79] Massey, W. S.: Singular homology theory. Berlin-Heidelberg-New York. Springer 1980.

[80] Metz, R.: Der affine Raum verallgemeinerter Reguli. Geom. Dedicata **10**, 337–367 (1981)

[81] Ostrom, T. G.: Derivable nets. Can. Math. Bull. **8**, 601–613 (1965)

[82] Percsy, N.: (p,q)-transitivité et configurations d' un plan projectif. Abh. Math. Sem. Univ. Hamb. **52**, 187–190 (1982)

[83] Pickert, G.: Analytische Geometrie. 2. Auflage. Leipzig 1955.

[84] Pickert, G.: Projektive Ebenen. 2. Auflage. Berlin-Heidelberg-New York. Springer 1975.

[85] Prieß-Crampe, S.: Angeordnete Strukturen: Gruppen, Körper, projektive Ebenen. Berlin-Heidelberg-New York-Tokyo. Springer 1983.

[86] Rees, D.: Nuclei of non-associative division algebras. Proc. Cambridge Philos. Soc. **46**, 1–18 (1950)

[87] Salzmann, H.: Topological planes. Adv. Math. **2**, 1–60 (1967)

[88] Salzmann, H.: Kollineationsgruppen kompakter, vier-dimensionaler Ebenen. Math. Z. **117**, 112–124 (1970)

[89] Salzmann, H.: Compact 8-dimensional projective planes with large collineation groups. Geom. Dedicata **8**, 139–161 (1979)

[90] Salzmann, H.: Automorphismengruppen 8-dimensionaler Ternärkörper. Math. Z. **166**, 265–275 (1979)

[91] Salzmann, H., Betten, D., Grundhöfer, Th., Hähl, H., Löwen, R. and Stroppel, M.: Compact projective planes. Berlin. De Gruyter 1995.

[92] Schafer, R. D.: Introduction to non-associative algebras. New York. Academic Press 1966.

[93] Schröder, E. M.: Darstellung der Gruppenräume Minkowskischer Ebenen. Arch. Math. **21**, 308–316 (1970)

[94] Schröder, E. M.: Kennzeichnung und Darstellung kinematischer Räume metrischer Ebenen. Abh. Math. Sem. Univ. Hamb. **39**, 184–230 (1973)

[95] Schrutka von Rechtenstamm, L.: Über die Auflösung linearer Quaternionengleichungen. Sitz.-Ber. Ak. (math.-nat.) Wien **115**, 739–775 (1906)

[96] Segre, B.: Lectures on modern geometry. Roma. Ed. Cremonese 1962

[97] Sherk, F. A.: Indicator sets in an affine space of any dimension. Can. J. Math. **31**, 211–224 (1979)

[98] Sherk, F. A. and Pabst, G.: Indicator sets, reguli, and a new class of spreads. Can. J. Math. **29**, 132–154 (1977)

[99] Thas, J. A.: Flocks of non-singular ruled quadrics. Atti Accad. Nac. Lincei **59**, 83–85 (1975)

[100] Tits, J.: Buildings of spherical type and finite BN-pairs. Lecture Notes in Mathematics **386**. Berlin-Heidelberg-New York- Tokyo. Springer 1986

[101] Weber, H.: Lehrbuch der Algebra I. Braunschweig. Vieweg 1895.

# Index

# Springer-Verlag
# and the Environment

We at Springer-Verlag firmly believe that an international science publisher has a special obligation to the environment, and our corporate policies consistently reflect this conviction.

We also expect our business partners – paper mills, printers, packaging manufacturers, etc. – to commit themselves to using environmentally friendly materials and production processes.

The paper in this book is made from low- or no-chlorine pulp and is acid free, in conformance with international standards for paper permanency.

# Lecture Notes in Mathematics

For information about Vols. 1–1431
please contact your bookseller or Springer-Verlag

Vol. 1518: H. Stichtenoth, M. A. Tsfasman (Eds.), Coding Theory and Algebraic Geometry. Proceedings, 1991. VIII, 223 pages. 1992.

Vol. 1519: M. W. Short, The Primitive Soluble Permutation Groups of Degree less than 256. IX, 145 pages. 1992.

Vol. 1520: Yu. G. Borisovich, Yu. E. Gliklikh (Eds.), Global Analysis – Studies and Applications V. VII, 284 pages. 1992.

Vol. 1521: S. Busenberg, B. Forte, H. K. Kuiken, Mathematical Modelling of Industrial Process. Bari, 1990. Editors: V. Capasso, A. Fasano. VII, 162 pages. 1992.

Vol. 1522: J.-M. Delort, F. B. I. Transformation. VII, 101 pages. 1992.

Vol. 1523: W. Xue, Rings with Morita Duality. X, 168 pages. 1992.

Vol. 1524: M. Coste, L. Mahé, M.-F. Roy (Eds.), Real Algebraic Geometry. Proceedings, 1991. VIII, 418 pages. 1992.

Vol. 1525: C. Casacuberta, M. Castellet (Eds.), Mathematical Research Today and Tomorrow. VII, 112 pages. 1992.

Vol. 1526: J. Azéma, P. A. Meyer, M. Yor (Eds.), Séminaire de Probabilités XXVI. X, 633 pages. 1992.

Vol. 1527: M. I. Freidlin, J.-F. Le Gall, Ecole d'Eté de Probabilités de Saint-Flour XX – 1990. Editor: P. L. Hennequin. VIII, 244 pages. 1992.

Vol. 1528: G. Isac, Complementarity Problems. VI, 297 pages. 1992.

Vol. 1529: J. van Neerven, The Adjoint of a Semigroup of Linear Operators. X, 195 pages. 1992.

Vol. 1530: J. G. Heywood, K. Masuda, R. Rautmann, S. A. Solonnikov (Eds.), The Navier-Stokes Equations II – Theory and Numerical Methods. IX, 322 pages. 1992.

Vol. 1531: M. Stoer, Design of Survivable Networks. IV, 206 pages. 1992.

Vol. 1532: J. F. Colombeau, Multiplication of Distributions. X, 184 pages. 1992.

Vol. 1533: P. Jipsen, H. Rose, Varieties of Lattices. X, 162 pages. 1992.

Vol. 1534: C. Greither, Cyclic Galois Extensions of Commutative Rings. X, 145 pages. 1992.

Vol. 1535: A. B. Evans, Orthomorphism Graphs of Groups. VIII, 114 pages. 1992.

Vol. 1536: M. K. Kwong, A. Zettl, Norm Inequalities for Derivatives and Differences. VII, 150 pages. 1992.

Vol. 1537: P. Fitzpatrick, M. Martelli, J. Mawhin, R. Nussbaum, Topological Methods for Ordinary Differential Equations. Montecatini Terme, 1991. Editors: M. Furi, P. Zecca. VII, 218 pages. 1993.

Vol. 1538: P.-A. Meyer, Quantum Probability for Probabilists. X, 287 pages. 1993.

Vol. 1539: M. Coornaert, A. Papadopoulos, Symbolic Dynamics and Hyperbolic Groups. VIII, 138 pages. 1993.

Vol. 1540: H. Komatsu (Ed.), Functional Analysis and Related Topics, 1991. Proceedings. XXI, 413 pages. 1993.

Vol. 1541: D. A. Dawson, B. Maisonneuve, J. Spencer, Ecole d´ Eté de Probabilités de Saint-Flour XXI - 1991. Editor: P. L. Hennequin. VIII, 356 pages. 1993.

Vol. 1542: J. Fröhlich, Th. Kerler, Quantum Groups, Quantum Categories and Quantum Field Theory. VII, 431 pages. 1993.

Vol. 1543: A. L. Dontchev, T. Zolezzi, Well-Posed Optimization Problems. XII, 421 pages. 1993.

Vol. 1544: M. Schürmann, White Noise on Bialgebras. VII, 146 pages. 1993.

Vol. 1545: J. Morgan, K. O'Grady, Differential Topology of Complex Surfaces. VIII, 224 pages. 1993.

Vol. 1546: V. V. Kalashnikov, V. M. Zolotarev (Eds.), Stability Problems for Stochastic Models. Proceedings, 1991. VIII, 229 pages. 1993.

Vol. 1547: P. Harmand, D. Werner, W. Werner, M-ideals in Banach Spaces and Banach Algebras. VIII, 387 pages. 1993.

Vol. 1548: T. Urabe, Dynkin Graphs and Quadrilateral Singularities. VI, 233 pages. 1993.

Vol. 1549: G. Vainikko, Multidimensional Weakly Singular Integral Equations. XI, 159 pages. 1993.

Vol. 1550: A. A. Gonchar, E. B. Saff (Eds.), Methods of Approximation Theory in Complex Analysis and Mathematical Physics IV, 222 pages, 1993.

Vol. 1551: L. Arkeryd, P. L. Lions, P.A. Markowich, S.R. S. Varadhan. Nonequilibrium Problems in Many-Particle Systems. Montecatini, 1992. Editors: C. Cercignani, M. Pulvirenti. VII, 158 pages 1993.

Vol. 1552: J. Hilgert, K.-H. Neeb, Lie Semigroups and their Applications. XII, 315 pages. 1993.

Vol. 1553: J.-L- Colliot-Thélène, J. Kato, P. Vojta. Arithmetic Algebraic Geometry. Trento, 1991. Editor: E. Ballico. VII, 223 pages. 1993.

Vol. 1554: A. K. Lenstra, H. W. Lenstra, Jr. (Eds.), The Development of the Number Field Sieve. VIII, 131 pages. 1993.

Vol. 1555: O. Liess, Conical Refraction and Higher Microlocalization. X, 389 pages. 1993.

Vol. 1556: S. B. Kuksin, Nearly Integrable Infinite-Dimensional Hamiltonian Systems. XXVII, 101 pages. 1993.

Vol. 1557: J. Azéma, P. A. Meyer, M. Yor (Eds.), Séminaire de Probabilités XXVII. VI, 327 pages. 1993.

Vol. 1558: T. J. Bridges, J. E. Furter, Singularity Theory and Equivariant Symplectic Maps. VI, 226 pages. 1993.

Vol. 1559: V. G. Sprindžuk, Classical Diophantine Equations. XII, 228 pages. 1993.

Vol. 1560: T. Bartsch, Topological Methods for Variational Problems with Symmetries. X, 152 pages. 1993.

Vol. 1561: I. S. Molchanov, Limit Theorems for Unions of Random Closed Sets. X, 157 pages. 1993.

Vol. 1562: G. Harder, Eisensteinkohomologie und die Konstruktion gemischter Motive. XX, 184 pages. 1993.

Vol. 1563: E. Fabes, M. Fukushima, L. Gross, C. Kenig, M. Röckner, D. W. Stroock, Dirichlet Forms. Varenna, 1992. Editors: G. Dell'Antonio, U. Mosco. VII, 245 pages. 1993.

Vol. 1564: J. Jorgenson, S. Lang, Basic Analysis of Regularized Series and Products. IX, 122 pages. 1993.

Vol. 1565: L. Boutet de Monvel, C. De Concini, C. Procesi, P. Schapira, M. Vergne. D-modules, Representation Theory, and Quantum Groups. Venezia, 1992. Editors: G. Zampieri, A. D'Agnolo. VII, 217 pages. 1993.

Vol. 1566: B. Edixhoven, J.-H. Evertse (Eds.), Diophantine Approximation and Abelian Varieties. XIII, 127 pages. 1993.

Vol. 1567: R. L. Dobrushin, S. Kusuoka, Statistical Mechanics and Fractals. VII, 98 pages. 1993.

Vol. 1568: F. Weisz, Martingale Hardy Spaces and their Application in Fourier Analysis. VIII, 217 pages. 1994.

Vol. 1569: V. Totik, Weighted Approximation with Varying Weight. VI, 117 pages. 1994.

Vol. 1570: R. deLaubenfels, Existence Families, Functional Calculi and Evolution Equations. XV, 234 pages. 1994.

Vol. 1571: S. Yu. Pilyugin, The Space of Dynamical Systems with the C⁰-Topology. X, 188 pages. 1994.

Vol. 1572: L. Göttsche, Hilbert Schemes of Zero-Dimensional Subschemes of Smooth Varieties. IX, 196 pages. 1994.

Vol. 1573: V. P. Havin, N. K. Nikolski (Eds.), Linear and Complex Analysis – Problem Book 3 – Part I. XXII, 489 pages. 1994.

Vol. 1574: V. P. Havin, N. K. Nikolski (Eds.), Linear and Complex Analysis – Problem Book 3 – Part II. XXII, 507 pages. 1994.

Vol. 1575: M. Mitrea, Clifford Wavelets, Singular Integrals, and Hardy Spaces. XI, 116 pages. 1994.

Vol. 1576: K. Kitahara, Spaces of Approximating Functions with Haar-Like Conditions. X, 110 pages. 1994.

Vol. 1577: N. Obata, White Noise Calculus and Fock Space. X, 183 pages. 1994.

Vol. 1578: J. Bernstein, V. Lunts, Equivariant Sheaves and Functors. V, 139 pages. 1994.

Vol. 1579: N. Kazamaki, Continuous Exponential Martingales and *BMO*. VII, 91 pages. 1994.

Vol. 1580: M. Milman, Extrapolation and Optimal Decompositions with Applications to Analysis. XI, 161 pages. 1994.

Vol. 1581: D. Bakry, R. D. Gill, S. A. Molchanov, Lectures on Probability Theory. Editor: P. Bernard. VIII, 420 pages. 1994.

Vol. 1582: W. Balser, From Divergent Power Series to Analytic Functions. X, 108 pages. 1994.

Vol. 1583: J. Azéma, P. A. Meyer, M. Yor (Eds.), Séminaire de Probabilités XXVIII. VI, 334 pages. 1994.

Vol. 1584: M. Brokate, N. Kenmochi, I. Müller, J. F. Rodriguez, C. Verdi, Phase Transitions and Hysteresis. Montecatini Terme, 1993. Editor: A. Visintin. VII. 291 pages. 1994.

Vol. 1585: G. Frey (Ed.), On Artin's Conjecture for Odd 2-dimensional Representations. VIII, 148 pages. 1994.

Vol. 1586: R. Nillsen, Difference Spaces and Invariant Linear Forms. XII, 186 pages. 1994.

Vol. 1587: N. Xi, Representations of Affine Hecke Algebras. VIII, 137 pages. 1994.

Vol. 1588: C. Scheiderer, Real and Étale Cohomology. XXIV, 273 pages. 1994.

Vol. 1589: J. Bellissard, M. Degli Esposti, G. Forni, S. Graffi, S. Isola, J. N. Mather, Transition to Chaos in Classical and Quantum Mechanics. Montecatini Terme, 1991. Editor: S. Graffi. VII, 192 pages. 1994.

Vol. 1590: P. M. Soardi, Potential Theory on Infinite Networks. VIII, 187 pages. 1994.

Vol. 1591: M. Abate, G. Patrizio, Finsler Metrics – A Global Approach. IX, 180 pages. 1994.

Vol. 1592: K. W. Breitung, Asymptotic Approximations for Probability Integrals. IX, 146 pages. 1994.

Vol. 1593: J. Jorgenson & S. Lang, D. Goldfeld, Explicit Formulas for Regularized Products and Series. VIII, 154 pages. 1994.

Vol. 1594: M. Green, J. Murre, C. Voisin, Algebraic Cycles and Hodge Theory. Torino, 1993. Editors: A. Albano, F. Bardelli. VII, 275 pages. 1994.

Vol. 1595: R.D.M. Accola, Topics in the Theory of Riemann Surfaces. IX, 105 pages. 1994.

Vol. 1596: L. Heindorf, L. B. Shapiro, Nearly Projective Boolean Algebras. X, 202 pages. 1994.

Vol. 1597: B. Herzog, Kodaira-Spencer Maps in Local Algebra. XVII, 176 pages. 1994.

Vol. 1598: J. Berndt, F. Tricerri, L. Vanhecke, Generalized Heisenberg Groups and Damek-Ricci Harmonic Spaces. VIII, 125 pages. 1995.

Vol. 1599: K. Johannson, Topology and Combinatorics of 3-Manifolds. XVIII, 446 pages. 1995.

Vol. 1600: W. Narkiewicz, Polynomial Mappings. VII, 130 pages. 1995.

Vol. 1601: A. Pott, Finite Geometry and Character Theory. VII, 181 pages. 1995.

Vol. 1602: J. Winkelmann, The Classification of Three-dimensional Homogeneous Complex Manifolds. XI, 230 pages. 1995.

Vol. 1603: V. Ene, Real Functions – Current Topics. XIII, 310 pages. 1995.

Vol. 1604: A. Huber, Mixed Motives and their Realization in Derived Categories. XV, 207 pages. 1995.

Vol. 1605: L. B. Wahlbin, Superconvergence in Galerkin Finite Element Methods. XI, 166 pages. 1995.

Vol. 1606: P.-D. Liu, M. Qian, Smooth Ergodic Theory of Random Dynamical Systems. XI, 221 pages. 1995.

Vol. 1607: G. Schwarz, Hodge Decomposition – A Method for Solving Boundary Value Problems. VII, 155 pages. 1995.

Vol. 1608: P. Biane, R. Durrett, Lectures on Probability Theory. VII, 210 pages. 1995.

Vol. 1609: L. Arnold, C. Jones, K. Mischaikow, G. Raugel, Dynamical Systems. Montecatini Terme, 1994. Editor: R. Johnson. VIII, 329 pages. 1995.

Vol. 1610: A. S. Üstünel, An Introduction to Analysis on Wiener Space. X, 95 pages. 1995.

Vol. 1611: N. Knarr, Translation Planes. VI, 112 pages. 1995.